Franziska Bosshard

Mechanisms of Cell Damage during Solar Disinfection (SODIS)

Franziska Bosshard

Mechanisms of Cell Damage during Solar Disinfection (SODIS)

Drinking Water Treatment by Solar Light causes fast Die-off in Enteric Pathogens

Südwestdeutscher Verlag für Hochschulschriften

Impressum/Imprint (nur für Deutschland/only for Germany)
Bibliografische Information der Deutschen Nationalbibliothek: Die Deutsche Nationalbibliothek verzeichnet diese Publikation in der Deutschen Nationalbibliografie; detaillierte bibliografische Daten sind im Internet über http://dnb.d-nb.de abrufbar.
Alle in diesem Buch genannten Marken und Produktnamen unterliegen warenzeichen-, marken- oder patentrechtlichem Schutz bzw. sind Warenzeichen oder eingetragene Warenzeichen der jeweiligen Inhaber. Die Wiedergabe von Marken, Produktnamen, Gebrauchsnamen, Handelsnamen, Warenbezeichnungen u.s.w. in diesem Werk berechtigt auch ohne besondere Kennzeichnung nicht zu der Annahme, dass solche Namen im Sinne der Warenzeichen- und Markenschutzgesetzgebung als frei zu betrachten wären und daher von jedermann benutzt werden dürften.

Coverbild: www.ingimage.com

Verlag: Südwestdeutscher Verlag für Hochschulschriften GmbH & Co. KG
Heinrich-Böcking-Str. 6-8, 66121 Saarbrücken, Deutschland
Telefon +49 681 37 20 271-1, Telefax +49 681 37 20 271-0
Email: info@svh-verlag.de

Approved by: Zürich, ETH, Diss., 2010

Herstellung in Deutschland:
Schaltungsdienst Lange o.H.G., Berlin
Books on Demand GmbH, Norderstedt
Reha GmbH, Saarbrücken
Amazon Distribution GmbH, Leipzig
ISBN: 978-3-8381-1061-5

Imprint (only for USA, GB)
Bibliographic information published by the Deutsche Nationalbibliothek: The Deutsche Nationalbibliothek lists this publication in the Deutsche Nationalbibliografie; detailed bibliographic data are available in the Internet at http://dnb.d-nb.de.
Any brand names and product names mentioned in this book are subject to trademark, brand or patent protection and are trademarks or registered trademarks of their respective holders. The use of brand names, product names, common names, trade names, product descriptions etc. even without a particular marking in this works is in no way to be construed to mean that such names may be regarded as unrestricted in respect of trademark and brand protection legislation and could thus be used by anyone.

Cover image: www.ingimage.com

Publisher: Südwestdeutscher Verlag für Hochschulschriften GmbH & Co. KG
Heinrich-Böcking-Str. 6-8, 66121 Saarbrücken, Germany
Phone +49 681 37 20 271-1, Fax +49 681 37 20 271-0
Email: info@svh-verlag.de

Printed in the U.S.A.
Printed in the U.K. by (see last page)
ISBN: 978-3-8381-1061-5

What really makes science grow is new ideas, including false ideas.

Karl Popper

Table of contents

Summary

The availability of drinking water is a key health issue in developing countries. Over 1.2 billion people lack access to safe drinking water today. They are using contaminated surface water from rivers, ponds and unprotected wells. This causes a high frequency of diarrhoeal diseases, which are life-threatening, especially for children under the age of five. About 4000 young children die from diarrhoea every day. The World Health Organization (WHO) has recognized the problem and promotes low-cost and effective home-based water treatment methods. Solar disinfection (SODIS) is one of these methods. In many parts of the world the UVA part of sunlight can be employed for disinfection purposes: 6 hours of exposure of unsafe drinking water to the sun in plastic PET bottles results in an inactivation of enteric pathogens (bacteria, protozoa and viruses) of several orders of magnitude. Over 3 million people in 33 countries are currently using SODIS in daily life. Although the method is widely used in practice, the cellular mechanisms leading to bacterial death during solar UVA irradiation still leaves many open questions. This thesis addresses some of these questions.

Cultivation on solid agar substrates has been used to demonstrate the efficacy of SODIS for many different organisms. However, the validity of this traditional method has been questioned in recent years since it was shown that the effectiveness of disinfection processes has often been overestimated when tested on this basis. In this thesis, the disinfection process during SODIS was analyzed with the latest culture-independent methods (multi-parameter flow cytometry and others) at the single cell level for the two important enteric pathogens *Shigella flexneri* and *Salmonella typhimurium*. Parameters investigated included cellular ATP levels, efflux pump activity, glucose uptake ability, polarization and integrity of the cytoplasmic membrane. The sequential break-down of the measured viability indicators was very well comparable to earlier results from *Escherichia coli*, a fact that allowed us to use this organism as a model organism for further mechanistic studies in this thesis. All viability

Summary

indicators used here suggest that membrane functions play a very critical role in SODIS inactivation of bacterial cells and the respiratory chain of enteric bacteria was identified to be a likely target of sunlight and UVA irradiation. Furthermore, during dark storage after irradiation, the physiological state of the cells continued to deteriorate even in the absence of irradiation: apparently bacterial cells were unable to repair the damage. This strongly suggests that a relatively small light dose is already sufficient to irreversibly damage the cells and that storage of treated water in bottles after irradiation does not allow re-growth of inactivated bacterial cells. Moreover, it was shown that light dose reciprocity is an important issue when using simulated sunlight. At high irradiation intensities, light dose reciprocity failed and resulted in an overestimation of the effect whereas reciprocity applied well around natural sunlight intensity.

Since membrane functions seem to play a crucial role during SODIS and the respiratory chain was identified as a possible vital first target during UVA irradiation, enzymes of different cellular compartments were tested for their activity. Strikingly, most membrane enzymes were damaged much earlier than enzymes of the cytoplasm. The respiratory chain and the F_1F_0-ATPase were confirmed to be the very first targets on the way to cell death. Already slightly irradiated cells (after less than one hour of sunlight) were very much affected in their ability to keep up essential parts of the energy metabolism. Protein damages are therefore a likely cause for membrane dysfunction during UVA irradiation. The underlying mechanism most probably is the enhanced generation of reactive oxygen species (ROS) during irradiation in biological membranes via Fenton reactions.

A broader investigation of protein damages in irradiated cells was therefore performed. Using up-to-date molecular methods, the changes in the proteome of irradiated E. coli cells were analyzed. Proteins were confirmed as important targets during SODIS. Oxidative damage to specific proteins was detectable by an immunoblot method for carbonylated proteins at very low fluences already. A consequence of these oxidative damages was aggregation of proteins. An advanced semi-quantitative proteomic approach was used to analyze proteins affected by aggregation. Aggregation was shown to target structural proteins and enzymes of many different cellular pathways at fluences achieved with light exposition corresponding to about two hours of natural sunlight. Targets included vital cellular functions like the transcription and translation apparatus, transport systems, amino acid synthesis and degradation, respiration, ATP synthesis, glycolysis, the TCA cycle, chaperone functions and catalase. The protein damage pattern caused by SODIS strongly resembles the pattern caused by reactive oxygen stress. Hence, sunlight probably accelerates cellular senescence and leads to the inactivation and finally death of cells.

Zusammenfassung

Die Gesundheit der Bevölkerung in Entwicklungsländern hängt stark von der ausreichenden Versorgung mit Trinkwasser ab. Über 1.2 Milliarden Menschen haben heute keinen Zugang zu sicherem Trinkwasser. Sie benützen meistens Oberflächenwasser von fragwürdiger Qualität aus Flüssen, Teichen und ungeschützten Brunnen. Dies verursacht eine häufiges Auftreten von Durchfallerkrankungen, die vor allem für Kinder unter fünf Jahren lebensgefährlich sein können. Täglich sterben etwa 4000 Kinder in diesem Alter an Durchfall. Um diesen Problemen zu begegnen, fördert die Weltgesundheitsorganisation WHO günstige und effiziente Wasserbehandlungsmethoden für einzelne Haushalte. Dazu gehört auch die Solare Desinfektion (SODIS). Der UV-Anteil des Sonnenlichts wird dabei für Desinfektionszwecke ausgenützt. Krankheitserreger (Bakterien, Protozoen und Viren) im fragwürdigen Trinkwasser werden innerhalb von 6 Stunden in PET-Plastikflaschen an der Sonne um mehrere Zehnerpotenzen reduziert. SODIS wird heute von 3 Millionen Menschen in 33 Ländern täglich genutzt. Obwohl die Methode in der Praxis so grossen Anklang findet, weiss man noch wenig über die zellulären Inaktivierungsmechanismen während der UVA-Desinfektion. Diese Dissertation beantwortet einige Fragen, die in diesem Zusammenhang gestellt werden können.

Die klassische Kultivierungstechnik auf Agar-Nährböden hat gezeigt, dass SODIS für viele verschiedene Organismen eine zuverlässige Desinfektionswirkung hat. Diese traditionelle Methode ist in den letzten Jahren allerdings vermehrt in Kritik geraten, weil aufgrund ihrer Ergebnisse die Desinfektionswirkung verschiedener Wasserbehandlungsmethoden überschätzt wurde. Darum wurde der SODIS-Desinfektionsprozess mit den neusten kulturunabhängigen Methoden auf Einzelzellebene für die zwei wichtigen Krankheitserreger *Shigella flexneri* and *Salmonella typhimurium* analysiert. Die untersuchten Parameter umfassten den zellulärem ATP-Gehalt, die Aktivität der

Zusammenfassung

Effluxpumpen, die Fähigkeit zur Glukoseaufnahme, ein intaktes Membranpotential und die Integrität der zytoplasmatischen Membran. Die Abfolge der Inaktivierung dieser Vitalitäts-Indikatoren während der Desinfektion entsprach weitgehend früheren Ergebnissen für *Escherichia coli*. Deshalb wurde dieser Organismus als Modellorganismus für weitere mechanistische Untersuchungen verwendet. Die Ergebnisse weisen auch darauf hin, dass Membranfunktionen sehr wichtig für die Inaktivierung von Bakterien sind und dass die Atmungskette wahrscheinlich ein empfindliches Angriffsziel während UVA- und Sonnenlichtbestrahlung darstellt. Der physiologische Zustand von einmal bestrahlten Zellen verschlechterte sich ausserdem danach bei Lagerung im Dunkeln zusehends. Die Zellen waren nicht fähig, die ihnen zugefügten Schäden wieder zu reparieren. Dies weist darauf hin, dass schon eine relativ niedrige Lichtdosis ausreicht um Zellen irreversibel zu schädigen und dass in der Praxis die Lagerung von Flaschen nach der Bestrahlung das Wiederaufwachsen von SODIS-behandelten Keimen nicht erlaubt.

Da Membranfunktionen während SODIS eine wichtige Rolle spielen und die Atmungskette schon als mögliche primäre Schadstelle vermutet wurde, haben wir Enzyme aus verschiedenen Zellkompartimenten während der Bestrahlung getestet. Tatsächlich wurden Membranenzyme wesentlich früher geschädigt als zytoplasmatische Enzyme. Die Atmungskette und die F_1F_0-ATPase konnten als früheste Angriffspunkte während des Inaktivierungsprozesses bestätigt werden. Schon sehr schwach bestrahlten Zellen (nach weniger als einer halben Stunde Sonnenlicht) können wichtige Teile ihres Energiemetabolismus nicht mehr aufrechterhalten. Dies weist darauf hin, dass Proteinschäden der Grund für den Zusammenbruch verschiedener Membranfunktionen während UVA-Bestrahlung sind. Die Bestrahlung bewirkt dabei wahrscheinlich eine verstärkte Bildung von reaktiven Sauerstoffspezies über Fentonreaktionen, die dann Schäden an Membranproteinen verursachen.

In diesem Kontext hat sich eine genauere Untersuchung der Proteinschäden in bestrahlten Zellen aufgedrängt. Neuste molekulare Methoden wurden verwendet um Veränderungen von Proteinen in bestrahlten *E. coli* Zellen nachzuweisen. Dabei konnte bestätigt werden, dass sehr viele Proteine durch SODIS empfindlich geschädigt werden. Manche Proteine wurden spezifisch oxidiert, was mit einer Immunoblotmethode für carbonylierte Proteine nachgewiesen werden konnte. Als eine Folge der oxidativen Veränderungen können betroffene Proteine aggregieren. Eine semi-quantitative Messmethode wurde verwendet, um aggregierte Proteine zu finden. Aggregation wurde sowohl bei Strukturproteinen als auch bei verschiedenen wichtigen Enzymen nachgewiesen. Wichtige Lebensfunktionen der Zelle, wie Transkription und Translation, Transportsysteme, Aminosäuresynthese und -abbau, Atmung, ATP-Synthese, Glykolyse, der Krebszyklus, Chaperonfunktionen und Katalaseaktivität wurden dabei empfindlich getroffen. Der Proteinschaden durch UVA glich dabei verblüffend jenem, der durch oxidativen Stress verursacht wird. Somit liegt die Vermutung nahe, dass Sonnenlicht zelluläre Alterungsprozesse beschleunigt und dabei zu Inaktivierung und Tod der bestrahlten Zellen führt.

1   General introduction

The need for simple and effective drinking water treatment

Diarrhoea caused by enteric pathogens is a predominant issue among the poorest segments of the population in developing countries. About 4000 children under the age of five die from diarrhoea every day. Older children miss the chance to attend school and adults are unable to work during the frequent episodes of diarrhoea. Unsafe drinking water supply, inadequate sanitation and insufficient hygiene behavior are the factors causing the lions share (88%) of all diarrhoea cases (Graf *et al.*, 2008). In slums of big cities in developing countries, e.g. in the Kibera slum in Nairobi, the poorest segments of the population lack basic infrastructure and have no access to safe drinking water. Water supply is often unreliable and organized illegally. This creates the shockingly unfair situation that the poorest pay higher prices for water than people living in upper-class districts of the city, whereas the quality of the source water is highly questionable. To improve the situation regarding water-related illnesses, a way to obtain safe drinking water with a modest use of resources is necessary. Since drinking water is vulnerable to recontamination during transport and storage, domestic methods of water treatment should be preferred to centralized methods. Besides the usual ways of treating water, e.g. boiling and chlorination, solar disinfection (SODIS) is a low-cost and simple method to obtain safe water.

Bacterial enteric pathogens have the guts of warm-blooded animals and humans as their primary habitat and are, therefore, usually more sensitive to solar radiation than environmental strains that had the chance to adapt during evolution. The deleterious effect of solar irradiation on enteric bacterial pathogens can be used to improve microbiological drinking water quality by SODIS. In many parts of the world the UV part of sunlight can be employed for disinfection purposes: 6 hours of exposure of unsafe drinking water to the sun in plastic PET bottles results in a several log-inactivation of enteric pathogens (bacteria, protozoa and viruses). Over 3 million people in 33 countries are currently using SODIS in daily life (Fig. 1.1). The impact of SODIS on health has been documented in several epidemiological field studies, e.g., in India, where a total of 40% of diarrhoeal diseases and 50% of diarrhoea episodes were prevented by the use of SODIS (Rose et al., 2006). A number of other (field) studies focused on bacterial, protozoan and fungal pathogens (Khaengraeng & Reed, 2005; King et al., 2008; Lonnen et al., 2005; Malato et al., 2009; Martin-Dominguez et al., 2005; Sichel et al., 2007; Wegelin et al., 1994) and health impact (Conroy et al., 1996; Conroy et al., 1999; Conroy et al., 2001; Graf et al., 2008; Hobbins et al., 2003; Hobbins, 2004; Rose et al., 2006). Hence, SODIS is applied and working in practice. Nevertheless, the cellular mechanisms of bacterial die-off are hardly known.

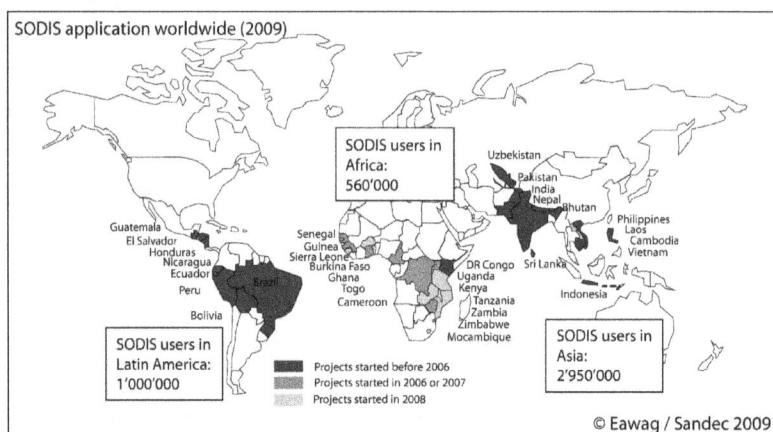

Fig. 1.1. SODIS application worldwide and starting dates of promotion projects.

Ultraviolet (UV) radiation in the solar light spectrum

Sunlight is an important environmental factor that has influenced life since cells have appeared on earth. In many surface ecosystems microorganisms are transiently affected by sunlight, a prominent example are the surface waters of the sea (Jeffrey *et al.*, 2005). The spectrum of sunlight consists of infrared, visible and ultraviolet (UV) radiation. UV light (190 - 400 nm) is divided into three wavelength ranges termed UVA (320 - 400 nm), UVB (290 - 320 nm) and UVC (190 - 290 nm) (Jagger, 1985). The atmosphere, especially the ozone layer in the stratosphere, filters out a large part of the sun's high frequency UV light (cut off at 300 nm). The intensity of UVA and UVB in sunlight reaching the earth's surface increases virtually linearly from 0 (at 300 nm) to roughly 0.5 W/m^2 (at 400 nm) on a mid-summer day (Berney *et al.*, 2006b). Integrated for the wavelength range 350 - 450 nm, the light dose (or "fluence") achieved during a mid-summer day in Europe is usually in the range of 2000 - 3000 kJ m^{-2}.

Proposed mechanisms of action
The pioneering work on the effect of solar light (UVA plus visible light of 350 - 490 nm) on bacteria was done in the 1940ies by Hollaender. He worked with the effects of "near"- and "far"-UV light on *E. coli* and already hypothesized that far-UV (he used 265 nm) directly damages nucleic acids, whereas near-UV (used was 350 - 490 nm) produces toxic compounds that destroy other cell components. Twenty years later, it was found that broad spectrum near-UV light can block the electron transport chain by photochemical decomposition of aromatic cofactors such as membrane-associated quinones (reviewed by (Jagger, 1972). Later, environmental engineers observed that sunlight significantly affects the survival of coliform bacteria shed into the environment with wastewater (Evison, 1988; Fujioka *et al.*, 1981; Gameson & Saxon, 1967; Kapuscinski & Mitchell, 1981; Kapuscinski, 1983). This effect was further investigated by Acra and coworkers for disinfecting oral rehydration solution and small quantities of drinking water (Acra, 1980; Acra, 1989) and his idea was further developed into SODIS by Wegelin and colleagues (Wegelin *et al.*, 1994).

Many times not only the effect of UV light but also that of mild heat (45 - 50°C, assumed to denature essential cellular enzymes) during exposures of SODIS flasks to the sun is assumed to be responsible for the inactivation of pathogens

3

(Malato *et al.*, 2009; Wegelin *et al.*, 1994). A synergy between temperature and near-UV radiation was observed already early (Hollaender, 1943; Tyrrell, 1976). However, to obtain sufficient synergetic effect, temperatures above 45 - 50°C are needed (Joyce *et al.*, 1996; Wegelin *et al.*, 1994) and field experiments demonstrated that temperatures above 45°C are rarely reached during application of SODIS (Joyce *et al.*, 1996; Reed, 1997b; Reed *et al.*, 2000). This aspect of inactivation mechanisms therefore was not in the focus of our work and all experiments presented in this thesis were conducted at a temperature of 37°C.

In general, mechanistic studies on bacterial die-off during UVA and solar irradiation are scarce. Different inactivation mechanisms were proposed to be important, some based on direct absorption of light by the damaged molecule, e.g., for nucleic acids, others acting indirectly by excitation of photosensitizers that then damage molecules in their neighborhood, e.g., in membranes. Indirect damage mechanism all seem to be based on the formation of reactive oxygen species (ROS) during irradiation. Previous findings from literature on damage mechanisms are shortly summarized here:

Membranes. If ROS lead to major damage due to sunlight, membranes and membrane proteins should be a major target since they contain a lot of sequestered iron. Indeed, membrane dysfunction upon UVA irradiation was reported by a number of researchers (Chamberlain & Moss, 1987; Kelland *et al.*, 1983a; Kelland *et al.*, 1983b; Kelland *et al.*, 1984; Klamen & Tuveson, 1982; Leven *et al.*, 1990; Mandal & Chatterjee, 1980; Pizarro, 1995; Sammartano & Tuveson, 1987; Tuveson *et al.*, 1987). It was shown that quinones, permease systems, galactoside transport and metabolic energy consumption are affected by UVA radiation (Jagger, 1972; Koch *et al.*, 1976). Increasing permeability of membranes or decreasing uptake rates of substrates were reported and leakiness of liposomal membrane particles was a direct consequence of exposure to either UVA light or sunlight (Koch *et al.*, 1976; Kubitschek & Doyle, 1981; Mandal & Chatterjee, 1980; Pizarro & Orce, 1988). Membrane functions seem to play a crucial role during UVA irradiation. We know from earlier studies in our laboratory that a similar sequential inactivation pattern was found in different bacterial species like *E. coli*, *S. typhimurium* and *S. flexneri* (Berney *et al.*, 2006a; Bosshard *et al.*, 2009a) with total ATP contents and efflux pump activity lost early (> 500 kJ m^{-2} in *E. coli*), followed by a complete depolarization of the cytoplasmatic membrane and loss of culturability (> 1500 kJ m^{-2}), and finally permeabilisation of the membrane (> 2000 kJ m^{-2}).

Nucleic acids. Compared to UVC light that is known to be absorbed by DNA directly, resulting in a number of different photoproducts (Ravanat *et al.*, 2001), UVA light was found to damage tRNA. An unusual base occurring in the 8-position in 65% of tRNAs species of *E. coli*, 4-thiouridine (s^{4}U), was shown to absorb light (Jagger, 1985; Kramer *et al.*, 1988). The base 4-thiouridine absorbs light optimally at about 340 nm resulting in the production of a crosslink with a cytidine at position 13 in the tRNA (Favre *et al.*, 1985). These crosslinked tRNAs were found to be poor substrates for amino acid charging and, therefore, amino acid availability is lowered and causes a shift-down of net protein synthesis (stringent response) (Favre & Hajnsdorf, 1983; Oppezzo & Pizarro, 2001; Ramabhadran & Jagger, 1976). However, a quantitative contribution of these mechanisms to the overall inactivation process by UVA light has not yet been demonstrated.

Reactive oxygen species (ROS). It was demonstrated early that UVA light inactivation of microorganisms strongly depends on the oxygen content of the water (Eisenstark, 1970; Reed, 1997b; Webb & Lorenz, 1970) and H_2O_2 was proposed as the photoproduct that may damage cellular components (Ahmad, 1981; Hartman & Eisenstark, 1978; Hartman *et al.*, 1979; McCormick *et al.*, 1976; Webb & Brown, 1979). Gene expression studies with continuous cultures of *E. coli* showed that the OxyR and SoxRS systems were induced during UVA irradiation (Berney *et al.*, 2006c). These systems are responsible to fight toxic action of elevated concentrations of ROS within a cell. Further, irradiated cells that were incubated anaerobically showed better survival on agar plates than cells that were incubated aerobically (Berney *et al.*, 2006a). This indicated that oxygen and/or its reactive derivatives play a major role in the inactivation process. ROS seem to target a wide variety of different molecules, because oxidation can literally harm any organic molecule. Studies of Hoerter and colleagues (Hoerter *et al.*, 2005a; Hoerter *et al.*, 2005b; Merwald *et al.*, 2005) support the suggestion that UVA light mediates its biological effect primarily as a result of ROS generated in some way. Solar light can activate oxygen, e.g., by transition metals via the so-called "photo-Fenton reaction" ($Fe^{2+} + H_2O_2 \rightarrow Fe^{3+} + OH^{\bullet} + OH^-$). This is described in more detail later.

The crucial role of oxygen during UVA irradiation by formation of ROS

ROS have a strong deleterious effect on many cell components. ROS generated during SODIS seem to be a main reason for cell inactivation. Many authors share this opinion and therefore, the generation of ROS in the light and their effect on cells are reviewed here.

The ability of molecular oxygen for oxidation of biomolecules is both a blessing and a curse for the cell. Respiring organisms exploit the high electronegativity of molecular oxygen to generate a membrane potential for ATP generation, but the uncontrolled oxidation of biomolecules should be avoided. Molecular oxygen is a so-called bi-radical, because its unpaired electrons occupy two orbitals. As a consequence, oxygen can adopt two spin-paired electrons from conventional organic molecules. It can only abstract one electron at a time and this considerably limits the reactivity of oxygen. The affinity of molecular oxygen for the first electron is very low (-0.16 V) (Fig. 1.2). Superoxide originates from molecular oxygen when one single electron is abstracted from a donor and hydrogen peroxide results when molecular oxygen abstracts two electrons. In contrast to the charged superoxide, hydrogen peroxide is able to cross biological membranes at neutral pH. ROS are formed with the unintended help of redox enzymes who transfer a single electron to the "wrong" acceptor. These can be flavoenzymes (Fig. 1.3), like succinate dehydrogenase, fumarate reductase, NADH dehydrogenase I & II, glutamate synthase, lipoamide dehydrogenase, and sulfite reductase, or enzymes containing Fe-S clusters (Fig. 1.4), like aconitases A, B, fumarases A, B, serine dehydratase, threonine deaminase, isopropylmalate isomerise, 6-phosphogluconate dehydratase. The auto-oxidation of flavoenzymes is probably the predominant source of both O_2^- and H_2O_2. Fe-S clusters release Fe^{2+} when oxidized by superoxide (Fig. 1.4). This additional free iron probably enhances Fenton reactions within the cell (via equations 1 & 2). Fe-S clusters are often very well hidden within the protein and are not well accessible for superoxide. Most respiratory dehydrogenases, therefore, are protected against the deleterious effects of superoxide (Imlay, 2009).

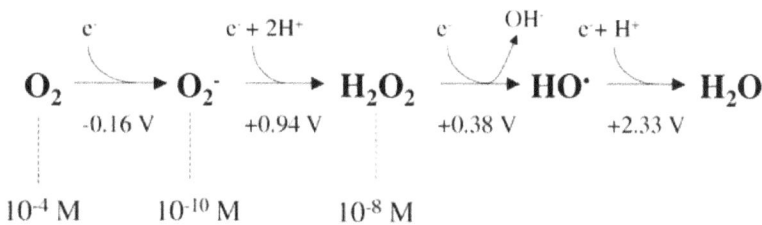

Fig. 1.2. The redox states of molecular oxygen. Left to right: molecular oxygen, superoxide, hydrogen peroxide, the hydroxyl radical, and water. Reduction potentials are shown; the reduction potential for molecular oxygen considers the standard state to be 1 M. At the bottom, the relative cytoplasmic concentrations of molecular oxygen, superoxide, and hydrogen peroxide in unstressed aerobic *E. coli* are shown (Imlay, 2009).

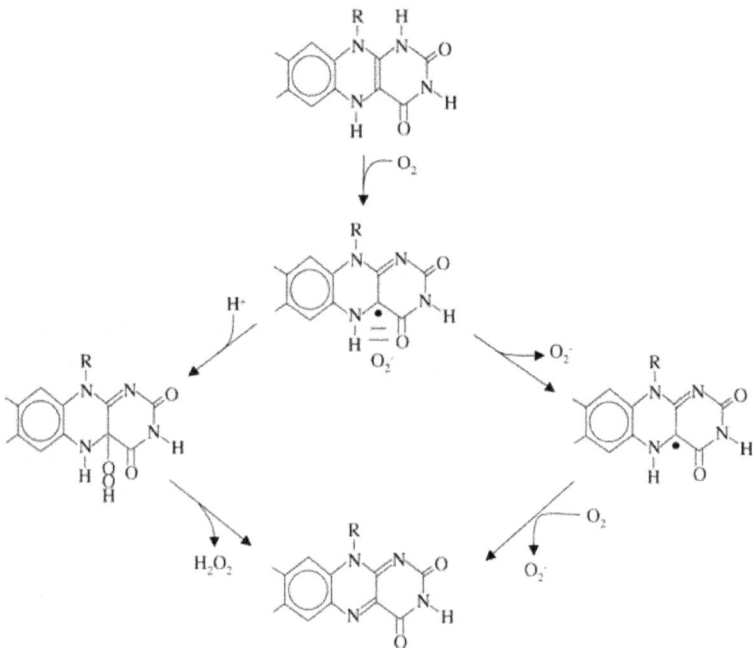

Fig. 1.3. The adventitious oxidation of enzymic dihydroflavin generates hydrogen peroxide and superoxide (Imlay, 2009).

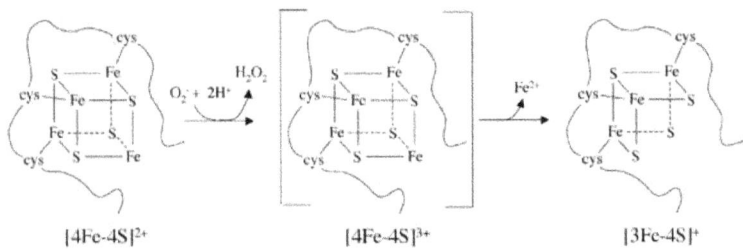

Fig. 1.4. Oxidative inactivation of a dehydratase [4Fe-4S] cluster by superoxide (Imlay, 2009).

The so-called "Fenton reaction" is responsible for the conversion of iron (II) and hydrogen peroxide to an iron (IV) intermediate (FeO^{2+}) and finally to an iron (III) and a highly aggressive hydroxyl radical.

$$Fe^{2+} + H_2O_2 \rightarrow (FeO^{2+}) \rightarrow Fe^{3+} + {}^{\bullet}OH \tag{1}$$

The hydroxyl radical is very reactive and, therefore, preferentially affects the neighborhood in which it was generated. Damages by hydroxyl radicals are thus expected everywhere near the sites where iron (II) is found within a cell, e.g. near Fe-S clusters or in enzymes using iron as prosthetic metal. Many studies indicate that iron is the primary source of hydroxyl radicals *in vivo* (Imlay, 2009).

The Fenton reaction is accelerated by light. Photo-reactive Fe (III) complexes absorb light at a wavelength of 436 nm and consequently, the iron photo-reduces to Fe (II).

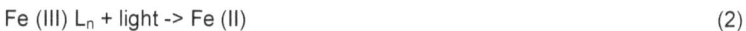

$$Fe\ (III)\ L_n + light \rightarrow Fe\ (II) \tag{2}$$

Like this, the iron is available for another round of ${}^{\bullet}OH$ production from H_2O_2 by the Fenton reaction shown in equation (1). Iron can be present in water as free iron or complexed to ligands such as oxalate, citrate, or phosphate. The so called "photo-Fenton reaction" works with all of these complexes efficiently in water at pH values ranging from 3 and 8 to produce ${}^{\bullet}OH$ radicals from H_2O_2. As an alternative, photolysis of H_2O_2 produces ${}^{\bullet}OH$ directly, but this process is slow compared to the photo-Fenton reaction since H_2O_2 only weakly absorbs sunlight (Zepp *et al.*, 1992).

Proteins containing flavins and FeS-centers are likely candidates to generate a big part of the cells' ROS load and, by this, also to be a major target of ROS.

Therefore, our hypothesis is that membrane proteins of the respiratory chain might be especially sensitive to oxidation during SODIS.

Protein carbonylation damage as a very probable consequence of increased ROS formation

Hydrogen peroxide and superoxide theoretically are able to oxidize amino acids directly, but these reactions are very slow compared to Fenton reactions acting via hydroxyl radicals and, therefore, do not have much physiological significance. Fenton reactions happening at the surface of proteins can either oxidize the α-carbon of the polypeptide backbone or attack amino acid side chains at high rates. Enzymes that use iron as a prosthetic metal were observed to be readily oxidizable by hydrogen peroxide *in vitro*, e.g., alcohol dehydrogenase (Tamarit *et al.*, 1998b) or the iron-containing superoxide dismutase (Beyer & Fridovich, 2002). Some enzymes that only loosely bind iron to the surface were inactivated efficiently by Fenton chemistry *in vitro* (Murakami *et al.*, 2006). Amino acid oxidation by a hydroxyl radical immediately leads to formation of another radical that can itself again react with nearby residues and thereby initiate a chain reaction. Several side-chain radicals are ultimately decomposed to carbonyl products (Fig. 1.5) (Imlay, 2009; Requena *et al.*, 2001; Requena *et al.*, 2003). Carbonyls on proteins are relatively easily detected by an immunoblot method initially developed by Levine and co-workers in the 90ies (Levine *et al.*, 1994; Shacter *et al.*, 1994) (Fig. 1.6). This assay has been widely used in medical research thereafter. A high level of carbonyl residues in proteins often accompanies senescence and disease. Even in prokaryotes it was found that increased protein carbonylation is age dependent. There are several reasons that could account for an increased carbonylation level in cells: a decline in the antioxidant defence system, increased ROS production, decreased capacity to remove oxidized proteins and a higher susceptibility of some proteins to oxidative attack. Also the availability of free iron in a cell can have an influence on the ROS level and result in increased carbonylation (Nyström, 2005).

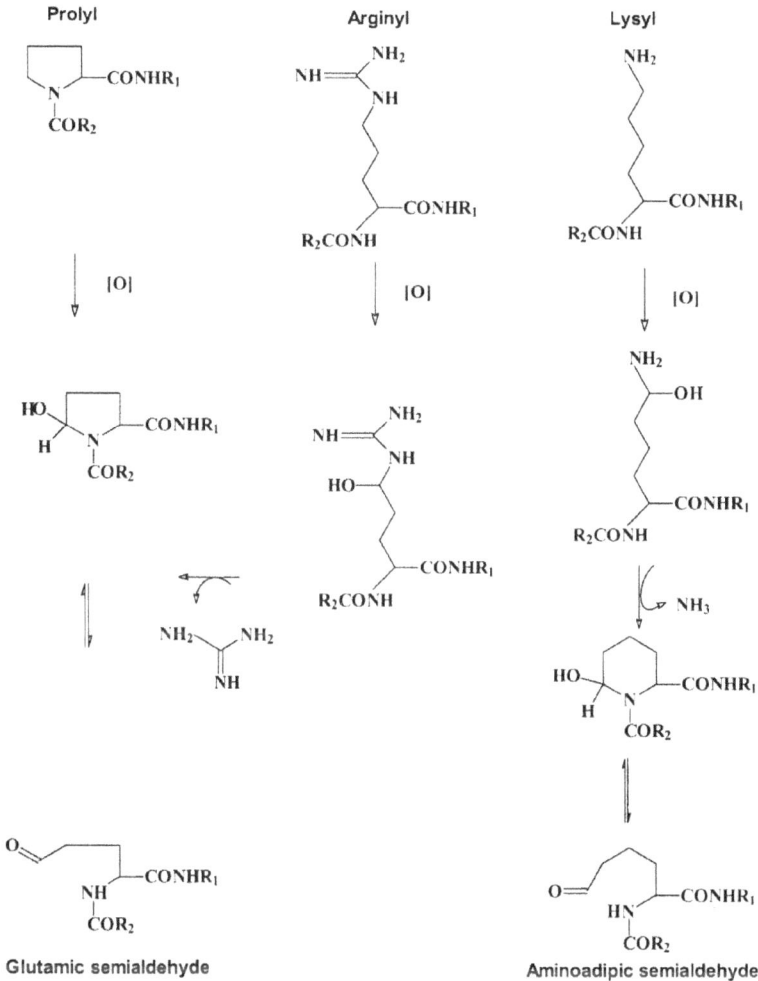

Fig. 1.5. Generation of glutamic and aminoadipic semialdehydes through oxidation of protein prolyl, arginyl and lysyl residues (Requena *et al.*, 2003).

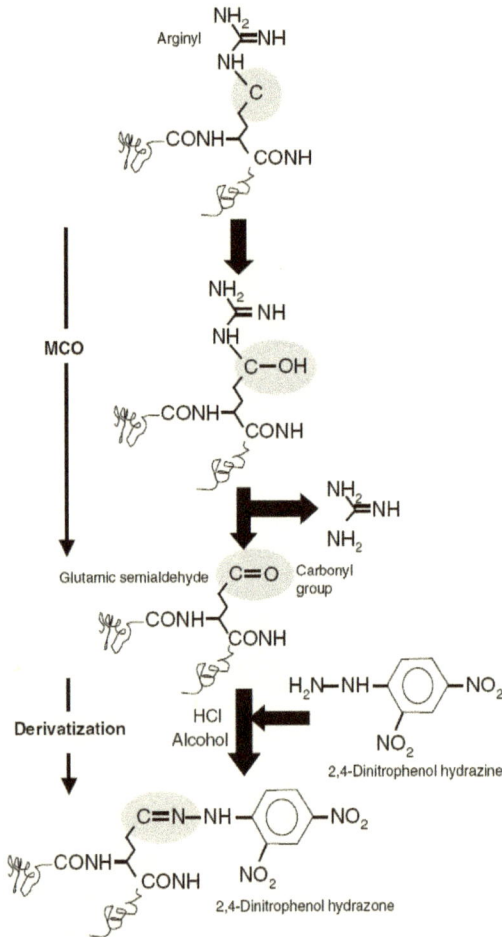

Fig. 1.6. Carbonylation and derivatization of a protein amino-acid side chain. As a consequence of a metal catalyzed oxidation (MCO) glutamic semialdehyde is derived from an arginyl residue. The resulting carbonyl group, in this case glutamic semialdehyde, is derivatized by 2,4-dinitrophenol hydrazine for detection. The protein 2,4-dinitrophenol hydrazone subsequently can be detected by specific monoclonal or polyclonal antibodies (Nyström, 2005).

Underlying hyptheses

We hypothesize that SODIS inactivates bacteria with the help of ROS, generated mostly in membranes via photo-Fenton reactions. This results in protein oxidation damages that target structural proteins and inactivate important enzymatic functions, first in membranes, then also in other parts of the cell. Oxidation of proteins leads also to protein fragmentation and accumulation of protein aggregates in irradiated cells. Since energy metabolism of cells is strongly dependent on intact membranes, the breakdown of membrane functions is suggested to be the primary cause for cellular die-off during SODIS.

Objectives of this thesis

In SODIS research, a lot of questions concerning characterization of cell damage during UVA irradiation are still unanswered, since mechanistic studies on bacterial viability and cell damage are scarce. The following research questions were addressed in this thesis:

Viability of pathogens during SODIS

- Do the membrane-associated inactivation patterns shown for *E. coli* and flow cytometric viability staining (Berney *et al.*, 2006a) translate to enteric pathogens?

Damages to proteins and enzymatic functions

- Are increased levels of oxidized proteins detected in irradiated cells? Are the patterns found similar to senescence patterns, does SODIS lead to accelerated senescence in irradiated cells?
- Are there other signs of protein and enzyme inactivation (inactivation of enzymes, protein aggregation and protein fragmentation)?
- Are membrane-associated enzymes early targets in the inactivation process compared to cytoplasmatic enzymes, since they are close to the most effective ROS generation sites within the cell?

Chapter 1

Dark storage after irradiation

- Are irradiated cells able to repair the damage that has occured? Is further cell inactivation in the dark taking place past irradiation? Is there a "point of no return", after which the damages in the cell are lethal during further dark storage?

Glossary

Fluence:	Integrated intensity during a specific time period (kJ/m^2) (SI units)
Fluence-rate:	Light intensity (W/m^2) (SI units)
Dose:	Synonym for fluence
Intensity:	Synonym for fluence-rate
Reciprocity:	states that total applied fluence produces the same response regardless of the fluence-rate
UVA:	Ultraviolet A, wavelength range 320 - 400 nm, synonym for near-UV or NUV
UVB:	Ultraviolet B, wavelength range 290 - 320 nm, synonym for mid-UV or MUV
UVC:	Ultraviolet C, wavelength range 190 - 290 nm, synonym for far-UV or FUV
ROS:	Reactive oxygen species: H_2O_2, $OH^{\bullet}$, O_2^-, $^1O_2$

2 Solar disinfection (SODIS) and subsequent dark storage of *Salmonella typhimurium* and *Shigella flexneri* monitored by flow cytometry

Abstract

Pathogenic enteric bacteria are a main cause of drinking water related morbidity and mortality in developing countries. Solar disinfection (SODIS) is an effective means to fight this problem. In the present study, solar disinfection of two important enteric pathogens, *Shigella flexneri* and *Salmonella typhimurium,* was investigated with a variety of viability indicators including cellular ATP levels, efflux pump activity, glucose uptake ability, polarisation and integrity of the cytoplasmic membrane. The respiratory chain of enteric bacteria was identified to be a likely target of sunlight and UVA irradiation. Furthermore, during dark storage after irradiation, the physiological state of the cells continued to deteriorate even in the absence of irradiation: apparently bacterial cells were unable to repair damage. This strongly suggests that for *Salmonella typhimurium* and *Shigella flexneri*, a relatively small light dose is enough to irreversibly damage the cells and that storage of bottles after irradiation does not allow regrowth of inactivated bacterial cells. In addition, we show that light dose reciprocity is an important issue when using simulated sunlight. At high irradiation intensities (> 700 W m^{-1}) light dose reciprocity failed and resulted in an overestimation of the effect whereas reciprocity applied well around natural sunlight intensity (< 400 W m^{-1}).

This chapter has been published in *Microbiology* **155**, p1310-1317 by F. Bosshard, M. Berney, M. Scheifele, H.-U. Weilenmann and T. Egli

Chapter 2

Introduction

The availability of safe drinking water is a key health issue in developing countries. The United Nations have declared it a millennium development goal to reduce the number of people without sustainable access to safe drinking water by half until 2015. Solar disinfection (SODIS) is one of the means to reach this goal. Its success is based on easy-available and low-cost tools: one day of exposure to the sun of hygienically unsafe drinking water in PET bottles leads to a significant increase in microbiological water quality. A positive impact on health has been documented in several epidemiological field studies, e.g. in India, where a total of 40% of diarrhoeal diseases and 50% of severe diarrhoea episodes were prevented by the use of SODIS (Rose et al., 2006). SODIS water treatment is already used by 2 million people and the number is increasing. But despite the fact that the method is working, the exact mechanism of inactivation of microbial pathogens is not yet known. Therefore, it is crucial to understand the way SODIS damages bacteria and whether repair can occur.

The effectiveness of SODIS has been proven by cultivation-based techniques with *Escherichia coli* and some pathogenic organisms (Acra, 1980; Berney et al., 2006b; McGuigan et al., 1998; Wegelin et al., 1994), and recently we have applied cultivation-independent methods to characterize the inactivation of *E. coli* by sunlight (Berney et al., 2006a). In that study we used flow cytometry combined with viability staining to characterize the loss of essential cellular functions in irradiated bacterial cells. The recorded cellular functions include membrane integrity, membrane potential, efflux pump activity and glucose uptake activity. We showed that a reproducible sequence of membrane-function breakdown takes place when *E. coli* is irradiated with sunlight or UVA light (Berney et al., 2006a). However, it is important to know whether these results translate to enteric pathogens like *Salmonella* or *Shigella,* the inactivation of which is the primary goal of solar disinfection. So far this has not been investigated.

18

Reliability of the SODIS method depends not only on the light dose leading to damage in target cells, but also on possible recovery processes in injured cells after irradiation. So far, no regrowth or recovery of membrane functions in injured *E. coli* cells was found (Berney *et al.*, 2006a; Joyce *et al.*, 1996; Oates *et al.*, 2003; Reed, 1997a; Wegelin *et al.*, 1994).

The present work extends our knowledge about the inactivation mechanism of solar light from the indicator bacterium *E. coli* to the two important enteric pathogens *Salmonella typhimurium* and *Shigella flexneri*. In addition, we investigated the ability of these enteric pathogens to survive and repair damage after solar irradiation.

Chapter 2

Methods

Bacterial strains. *Salmonella enterica* serovar Typhimurium ATCC 14028 (referred to in this paper as *Salmonella typhimurium*) and *Shigella flexneri* ATCC 12022 were used in this study.

Growth media and cultivation conditions. Cells were grown as described by (Berney *et al.*, 2006a) with modifications. In drinking water, cells grow very slowly or not at all; therefore, we used stationary-phase cells, which were shown to be more resistant to SODIS than cells in the exponential growth phase (Berney *et al.*, 2006b; Reed, 1997b). Luria-Bertani (LB) broth, which was filter-sterilized with membrane filters (0.22 µm, Millipore, Ireland) and diluted to 33% (v/v) of its original strength with ultrapure water, was used for batch cultivation. Precultures were prepared for each individual batch experiment from the same cryo-vial streaking the stock solution onto Hektoen agar plates (Oxoid, Hampshire, England) selective for *Shigella* and *Salmonella* species. After 15 - 18 h of incubation at 37 °C, one colony was picked, loop-inoculated into a 125 ml Erlenmeyer flask containing 20 ml of LB broth, and incubated at 37 °C on a rotary shaker at 200 r.p.m.. At an OD_{546} between 0.1 and 0.2, an aliquot of the culture appropriate to obtain an initial OD_{546} of 0.002 was transferred into a 500 ml Erlenmeyer flask containing 50 ml of prewarmed LB broth. With this procedure, no lag phase was observed. These flasks were then shaken at 200 r.p.m. on a rotary shaker at 37 °C for approximately 18 h until stationary phase was reached. Stationary phase was confirmed from five consecutive OD_{546} measurements within one hour.

Sample preparation and plating. Cells were harvested by centrifugation (16000g, 3min) from batch culture, washed three times with filter-sterilized commercially available bottled water (Evian, France) and diluted to an OD_{546} of approximately 0.01 (corresponding to 1-5 x 10^7 cells ml^{-1}). To allow the cells to adapt to the mineral water, light exposure of bacterial suspensions was started

20

only 1 h after dilution. During exposure, aliquots were withdrawn at different time points and diluted in decimal steps (10^{-1} to 10^{-6}) with sterile-filtered, bottled mineral water (Evian, France). Volumes of 1 ml of appropriate dilutions were withdrawn and mixed with 7 ml liquid tryptic soy agar (TSA) (Biolife, Milano, Italy) at 45 °C (pour-plate method). After 20 minutes, the solidified agar was covered with another 4 ml of liquid TSA (40 °C). Plates were incubated for 48 h at 37 °C until further analysis. The standard error of pour plating was always < 10 %.

Sunlight and UVA exposure. Samples of 10 ml bacterial suspension (see above) were exposed to sunlight or UVA light as described earlier (Berney *et al.*, 2006a; Berney *et al.*, 2006b; Berney *et al.*, 2007a).

Dark storage. Samples of 10 ml bacterial suspension were exposed to UVA light (see above). Cellular damages were assessed right after irradiation and at different time points during dark storage, which was performed at 37 °C for 48 h, holding the cells in the same medium (sterile-filtered bottled water) as during irradiation. To exclude the possibility of regrowth, nalidixic acid was included in all of the samples at a concentration of 100 µg ml^{-1}.

Staining procedures. Five fluorescent dyes were used alone or in different combinations: Syto 9 (Invitrogen Molecular Probes, Oregon, USA), propidium iodide (PI; Invitrogen), bis-(1,3-dibutylbarbituricacid)trimethine oxonol (DiBAC$_4$(3); Invitrogen), ethidium bromide (EB; Fluka, Switzerland) and 2-[N-(7-nitrobenz-2-oxa-1,3-diazol-4-yl)amino]-2-deoxy-D-glucose (2-NBDG; Invitrogen). Samples taken from irradiation experiments (sunlight and artificial UVA) were divided into five subsamples and immediately stained with two mixtures of fluorescent dyes (Syto 9/PI and Syto 9/EB) and three single fluorescent dyes [DiBAC$_4$(3), Syto 9 and 2-NBDG]. Samples were incubated in the dark at 37 °C for 5 min (2-NBDG) or at room temperature for 10 min [DiBAC$_4$(3)], 15 min (Syto 9/EB), 20 min (Syto 9/PI) and 25 min (Syto 9), respectively, before analysis. Prior to flow-cytometric analysis, samples (~1 - 5 x 10^7 cells ml^{-1}) were diluted with sterile-filtered bottled water (Evian) to 1% (v/v) of the initial cell concentration (~1 - 5 x 10^5 cells ml^{-1} final concentration). Stock solutions of the dyes were prepared as follows: PI and Syto 9 were used from the LIVE/DEAD BacLight kit (Invitrogen), EB was prepared at 25 mM in distilled and filtered water, DiBAC$_4$(3) was prepared in dimethylsulfoxide (DMSO) at 10 mM, and 2-NBDG was dissolved in distilled and filtered water at 5 mM. All stock solutions were stored at -20°C. The working concentrations of Syto 9, PI, EB, DiBAC$_4$(3) and 2-NBDG were 5, 30, 30, 10 and 5 mM, respectively. 2-NBDG was added in combination with 2,4-dinitrophenol (final concentration 2 mM) (Natarajan & Srienc, 2000). At the beginning of each experiment a sample was incubated at 90°C for 3 minutes (in a 2 ml Eppendorf tube) as a control measurement for inactive bacteria. By comparing the staining pattern of heat-inactivated with untreated samples, electronic gates were set to differentiate negatively- and positively-stained populations (Berney et al., 2008).

Flow-cytometric measurements. The methods used here have been described recently (Berney et al., 2006a; Berney et al., 2007a). Flow-cytometric measurements were made using a Partec Cyflow space flow cytometer with 488 nm excitation from an argon ion laser running at 50 mW or 200 mW (for the fluorescent glucose analogue 2-NBDG), respectively. Green fluorescence was collected in the FL1 channel (520±20 nm), and red fluorescence in the FL3 channel (>590 nm) and all data were processed with the Flowmax software (Partec), and electronic gating with the software was used to separate positive signals from noise. The specific instrumental gain settings for these measurements were as follows: FL1 = 490, FL3 = 600, speed 3 (implying an event rate never exceeding 1000 events s^{-1}). All samples were collected as logarithmic (3 decades) signals and were triggered on the green fluorescence channel (FL1). The routine check of flow cytometer was performed every day for correct alignment with an FITC bead standard. This ensured accuracy in counting (volumetric counting device) and measured fluorescence intensities. Microscopic observation was performed on an Olympus BX50 microscope (Olympus Schweiz AG, Volketswil, Switzerland) equipped with filters HQ-F41-007 for PI and EB and

HQ-F41-001 for Syto 9, DiBAC$_4$(3) and 2-NBDG (all from AF Analysetechnik Tübingen, Germany).

Total ATP. For the determination of total ATP, the BacTiter-Glo system (Promega, Madison, Wisconsin, USA) was used. The BacTiter-Glo buffer was mixed with the lyophilized BacTiter-Glo substrate and equilibrated at room temperature. The mixture was stored over night at room temperature to ensure that all ATP was hydrolysed („burned off") and the background signal had decreased. A cell suspension of 100 µl was mixed in a 2 ml Eppendorf tube with an equal volume of the previously prepared BacTiter-Glo reagent (stored on ice). The sample was then briefly mixed by once pipetting up and down and put into a water bath at 37 °C for 30 seconds. The luminescence of the sample was measured in a luminometer (model TD-20/20; Turner BioSystems) immediately after incubation. A calibration curve with dilutions of pure rATP (Promega, P1132) was measured for each batch of BacTiter-Glo buffer. ATP concentration per cell was then calculated using this calibration curve and the total count measurements (Syto 9) from flow cytometry.

Reproducibility. All field experiments were conducted in three biological replicates on three different days. Irradiation intensity data were obtained from a weather station, which is located 300m away from the exposure site (BAFU/NABEL, EMPA Duebendorf, Switzerland). Sunlight intensity does vary with the weather conditions and therefore, the light dose at the sampling points was never exactly the same. Therefore, we show data from representative result.

Chapter 2

Results

Artificial UVA light exposure. The susceptibility of different properties of *S. typhimurium* and *S. flexneri* to artificial UVA light is listed in Tab. 2.1. Both organisms were less susceptible to UVA light than the indicator bacterium *E. coli* (Berney et al., 2006a). For example, when assessed with propidium iodide staining, the light dose needed for membrane permeabilisation of *S. typhimurium* was approximately three times, and for *S. flexneri* approximately two times higher than for *E. coli*. However, for all enteric bacteria, the same sequential inactivation pattern was observed with the measured viability indicators.

Because of this higher resistance, some of the laboratory experiments were conducted with much higher irradiation intensities than those normally achieved with natural sunlight. Typically, the maximum natural sunlight intensity at noon on mid-European longitude is about 130 W m^{-2} (integrated for the wavelength spectrum of 350 – 450 nm), whereas in this study, intensities between 163 W m^{-2} and 1315 W m^{-2} were used for artificial UVA exposure to achieve high enough fluences within a reasonable time period. We observed that, as soon as natural irradiation conditions were exceeded two- to threefold, the reciprocity law started to fail not only for the culturability of *S. flexneri* (Fig. 2.1) but also for all other measured viability indicators (data not shown). This clearly demonstrates that for these enteric pathogens high irradiation intensities in laboratory experiments results in an overestimation of the effect in comparison with the same light dose under natural sunlight conditions and that such data therefore are of limited use. Therefore, doses listed in Tab. 2.1 were obtained exclusively from experiments performed at light intensities that were in a range of natural conditions.

The difference in the fluence needed to achieve a three-log reduction when exposing the cells to different light intensities was a result of two effects. Firstly, the shape of the inactivation curve followed the "log-linear with a shoulder" model for samples irradiated with natural sunlight or artificial UVA of a corresponding intensity. However, the shoulder disappeared when cells were irradiated with very high intensities (919 W m^{-2} and 1315 W m^{-2} in Fig. 2.1). Secondly, the inactivation was three times faster with very high intensities as compared to natural conditions (slope of -0.0031 for intensities around 360 W m^{-2} vs. -0.0011 with intensities around 1100 W m^{-2}).

In general, the reciprocity law was valid for S. flexneri as long as intensities not exceeding 400 W m^{-2} were applied. Experiments conducted in this irradiation intensity range compared well with natural conditions. Reciprocity for S. typhimurium was given in a much wider range (from 50 to 700 W m^{-2}). However, when very high irradiation intensities (1000 W m^{-2}) were applied, the shoulder of the inactivation curve was shortened (data not shown).

Sunlight exposure. A decrease in culturability of more than 99% was observed (Fig. 2.2a, 2.3a) for S. typhimurium and S. flexneri during one day of solar irradiation (8 hours; 2300 and 2500 kJ m^{-2}, respectively). Efflux pump activity and ATP concentration had reached their lowest level by the end of the day (Fig. 2.2b,e and 2.3b,e). In contrast, inactivation of glucose uptake ability, the loss of membrane potential, and loss of membrane integrity was not observed with S. typhimurium and S. flexneri during one day of sunlight irradiation (Fig. 2.2f,c,d and 2.3f,c,d). Therefore, we decided to continue exposure on the next day. Interestingly, S. typhimurium and S. flexneri lost cellular activity during dark storage at 37 °C over night.

This is clearly shown for S. typhimurium, where an additional one-log reduction in colony-forming units (cfu) was observed after the night break (Fig. 2.2a). Accordingly, 60% of the cells lost their membrane potential during the night break and a slight reduction in membrane integrity was also observed (Fig. 2.2b). Furthermore, the percentage of cells able to take up glucose increased by 40% on the first day and was followed by a loss of 80% during the night break (Fig. 2.2f). ATP concentration and efflux pump activity reached their lowest level on the first day of irradiation (Fig. 2.2b,e). Interestingly, a more than twofold increase

25

in ATP concentration was observed initially before it rapidly decreased at approximately 1000 kJ m^{-2}. A similar increase in cellular activity with increasing exposure in the initial phase was observed also for the uptake of the fluorescent glucose analogue (see above).

Tab. 2.1. Comparision of *S. typhimurium* and *S. flexneri* susceptibilities to artificial UVA light and sunlight. Figures indicate the approximate light doses (+/- 20%) in kJ m^{-2} at which > 90% of the cells exhibited the properties indicated. Cells were exposed continuously to artificial UVA light with an intensity of 620 W m^{-2} for *S. typhimurium* and 360 W m^{-2} for *S. flexneri*. In the case of sunlight, cells were exposed during two subsequent days with a night break of approximately 12 h. Sunlight exposure on the first day reached 2300 kJ m^{-2} for *S. typhimurium* and 2500 kJ m^{-2} for *S. flexneri*, respectively. An asterisk (*) indicates that this parameter changed during the night break. For comparison, corresponding data for *E. coli* are shown (Berney *et al.*, 2006a).

	S. typhimurium		*S. flexneri*		*E. coli*	
	UVA	sunlight	UVA	sunlight	UVA	sunlight
ATP level decreased to < 10% compared to control at	2000	1500	1000	1000	300	700
Loss of culturability (0.1% survival)	3000	2300*	1800	1800	1700	1700
> 90% of cells with inactivated efflux pumps at	3000	2300*	2000	2000	1000	1700
> 90% of cells unable to take up glucose at	3500	2300*	3000	3000	2000	2000
> 90% of cells with depolarized membranes at	6000	2300*	4500	2500*	1900	1700
> 90% of cells with permeabilised membranes at	8000	4500	6000	4500	2400	not achieved

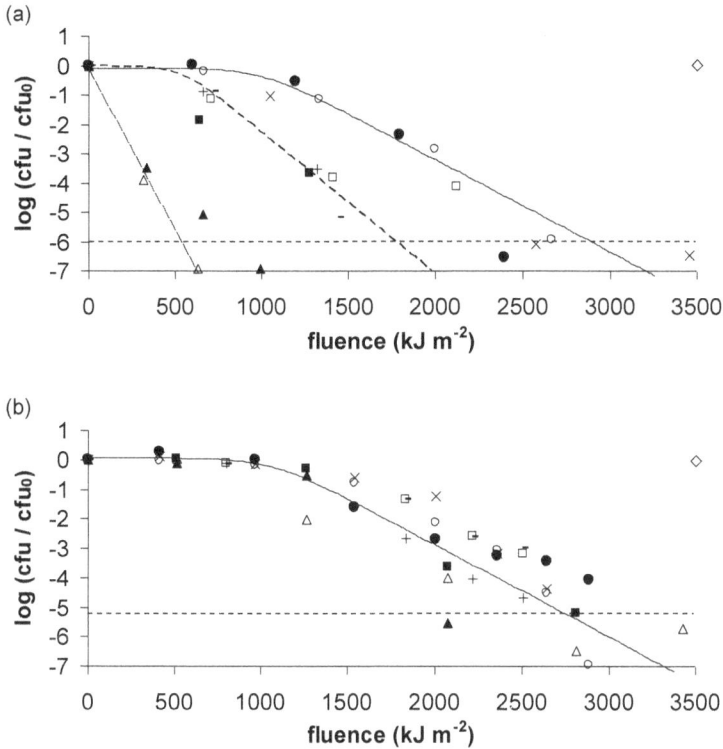

Fig. 2.1. Deviations from light dose reciprocity appearing with different intensities of artificial UVA light irradiation in comparison to long, low intensity irradiation with sunlight shown for the culturablity of *S. flexneri*. Bacterial cells were harvested from the stationary phase of a LB batch culture, washed three times and diluted in bottled mineral water. Cfu were measured by pour plating and sensitivity was recorded as cfu / (cfu at time zero). Dashed lines indicate the detection limits. Lines represent modelled inactivation curves with the program GInaFIT (Geeraerd *et al.*, 2005). Empty diamonds (◊) represent averaged controls. (a) Artificial UVA light was applied at an intensity of (X) 163 W m^{-2}, (●) 332 W m^{-2}, (○) 370 W m^{-2}, (■) 710 W m^{-2}, (+) 732 W m^{-2}, (□) 786 W m^{-2}, (-) 802 W m^{-2}, (▲) 919 W m^{-2} and (Δ) 1315 W m^{-2}. (b) Sunlight irradiation on three different days in biologically independent triplicates: (X), (●) and (○) on 08/23/2006; (■), (+) and (□) on 08/31/2006; (-), (▲) and (Δ) on 07/25/2006.

For *S. flexneri* a similar general pattern was observed, but with some distinct differences in the magnitude of the effects. Culturability dropped over more than 3 additional log-units during the night break (Fig. 2.3a). After the night break, 80% of the cells had lost their membrane potential and 20% were even permeabilised (Fig. 2.3c,d). About 40% of the cells were unable to take up glucose (Fig. 2.3f). As in *S. typhimurium*, ATP concentration and efflux pump activity in *S. flexneri* had reached already final levels on the first day. Comparable to *S. typhimurium*, a slight increase in glucose uptake activity was observed during irradiation on the first day (Fig. 2.3d).

The overall fluence rate resulting in membrane permeabilisation in > 90% of the cells (8000 vs. 4500 kJ m^{-2} in *S. typhimurium*, and 6000 vs. 4500 kJ m^{-2} in *S. flexneri*) and loss of membrane potential in > 90% of the cells (6000 vs. 2300kJ/m^2 in *S. typhimurium*, and 4500 vs. 2500 kJ m^{-2} in *S. flexneri*) was clearly reduced with discontinuous irradiation (Tab. 2.1).

Dark storage of *S. typhimurium* after irradiation. For a better understanding of the overnight processes in the outdoor experiments, irradiation and subsequent storage in the dark of *S. typhimurium* was investigated in the laboratory (Fig. 2.4). This was done because of the higher resistance of *S. typhimurium* resulting in only 2-log reduction during one full day of exposure. Bacterial cell suspensions were irradiated with a dose of 1500 kJ m^{-2} corresponding to half a day of sunlight or one full day of exposure at overcast conditions. Culturability, efflux pump activity, membrane potential, membrane permeability and ATP content per cell were measured right after exposure and for a period of up to 48 h at eight subsequent time points during storage in the dark. This aspect was investigated in more detail because it is possible that a small part of cells might survive the irradiation and that these survivors might start growing on either lysed cells or assimilable organic carbon (AOC) in the water. For example, it has been shown that some pathogenic bacteria are able to grow on natural substrates in bulk water (Vital *et al.*, 2007; Vital *et al.*, 2008). To exclude this possibility and to make sure that only the cells originally exposed to light were analyzed, we added nalidixic acid to inhibit cell division. The concentration of nalidixic acid required to keep *S. typhimurium* from dividing was determined in a minimum inhibitory concentration experiment.

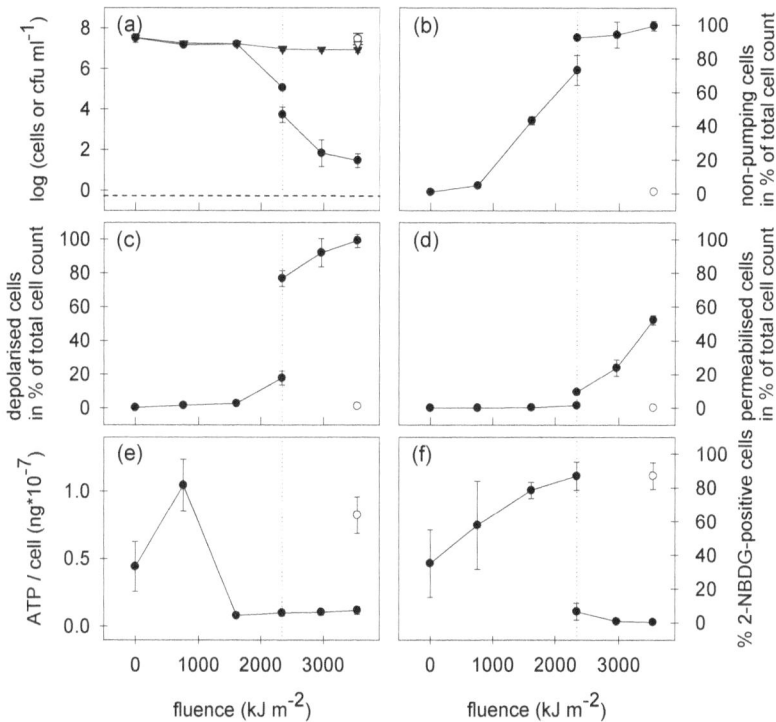

Fig. 2.2. Viability parameters exhibited by stationary phase cells of *S. typhimurium* exposed to sunlight. (a) culturability (●) log (cfu ml^{-1}) with the dashed line indicating the detection limit, (▼) log (total cell concentration ml^{-1}). (b) (●) EB-positive cells, (c) (●) DiBAC$_4$(3)-positive cells, (d) (●) PI-positive cells. Values were calculated as percentage (%) of total cell concentration. (e) (●) Average ATP concentration per cell. (f) (●) 2-NBDG-positive cells (able to take up glucose) calculated as percentage (%) of total cell concentration. Light dose on day 1: 2300 kJ m^{-2}, day 2: 1200 kJ m^{-2} (overcast conditions). The night break is indicated by a dotted line in each graph. In all graphs, unirradiated control

samples are displayed as empty symbols. Error bars represent standard deviations from 3 biologically independent experiments.

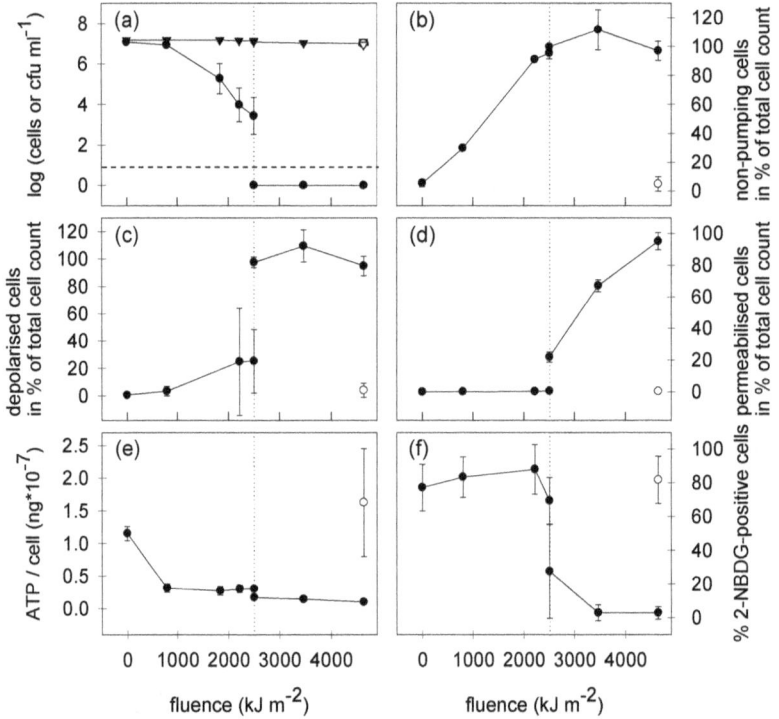

Fig. 2.3. Viability parameters exhibited by stationary phase cells of *S. flexneri* exposed to sunlight. (a) culturability ($\bullet$) log (cfu ml^{-1}) with the dashed line indicating the detection limit, ($\blacktriangledown$) log (total cell concentration ml^{-1}), (b) ($\bullet$) EB-positive cells, (c) ($\bullet$) DiBAC$_4$(3)-positive cells, (d) ($\bullet$) PI-positive cells. Values were calculated as percentage (%) of total cell concentration. (e) ($\bullet$) Average ATP concentration per cell. (f) ($\bullet$) 2-NBDG-positive cells (able to take up glucose) calculated as percentage (%) of total cell concentration. Light dose on day 1: 2500 kJ m^{-2}, day 2: 1300 kJ m^{-2} (overcast conditions). The night break is indicated by a dotted line in each graph. In all graphs, non-irradiated control samples are displayed as empty symbols. Error bars represent standard deviations from 3 biologically independent experiments.

After having received a „half-day"-UVA dose, cfu had decreased approximately 1 log-unit (Fig. 2.4a). During subsequent dark storage, cfu decreased by 5 log-units over 24 h. Efflux pump activity was lost completely just after irradiation and was not regained (Fig. 2.4b). The unirradiated control cells were also hampered in the beginning of the experiment, when about 50% of the population showed no efflux pump activity, probably as an effect of nalidixic acid. Efflux pump activity was regained later on in the control sample. Membrane potential was lost in only a very small part of the cell population during irradiation, but a relevant fraction of the cells (about 60%) lost their membrane potential in the following 24 h of dark storage (Fig. 2.4c).

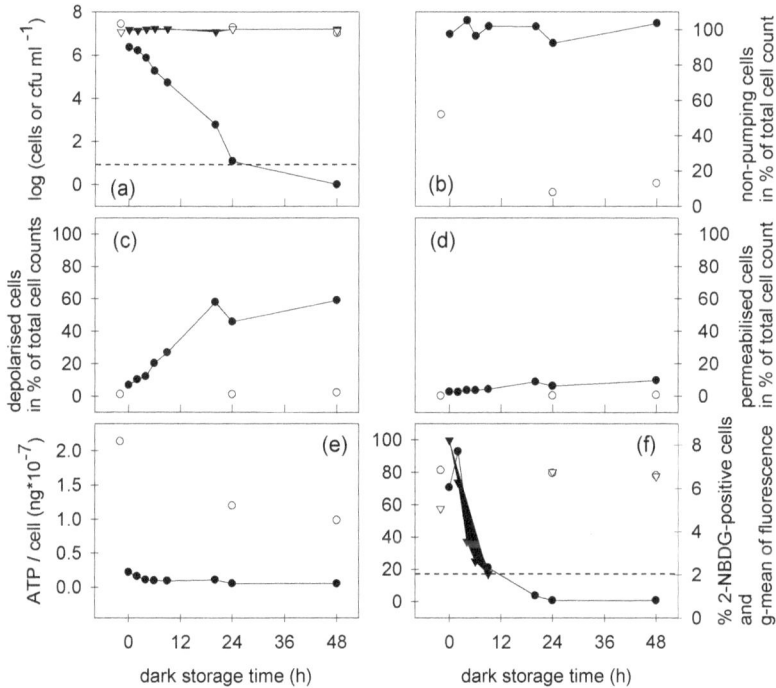

Fig. 2.4. Dark storage of *S. typhimurium* after exposure to a dose of 1500 kJ m⁻² artificial UVA light, applied with an irradiation intensity of 205 W m⁻² for 120 min.

In order to exclude the possibility of regrowth, nalidixic acid was included in all of the samples at a concentration of 100 µg ml^{-1}. (a) culturability (•) log (cfu ml^{-1}) with dashed line for detection limit, (▲) log (total cell concentration ml^{-1}). (b) (•) EB-positive cells, (c) (•) DiBAC$_4$(3)-positive cells, (d) (•) PI-positive cells. Values were calculated as percentage (%) of total cell concentration. (e) (•) Average ATP concentration per cell. (f) (•) 2-NBDG-positive cells (able to take up glucose) calculated as percentage (%) of total cell concentration with labels on left side of graph and (▼) geometrical mean of green fluorescence of the 2-NBDG-positive cell population with labels on right side of graph and dashed line for detection limit. In all graphs, non-irradiated control samples are displayed as empty symbols.

Integrity of the cell membranes did not change significantly during dark storage, only a slight increase in the percentage of permeabilised cells was observed (Fig. 2.4d). ATP content of the cells was reduced to 10% compared to the control just after the treatment and levelled off to < 5% during the first 10 h after the treatment. Also the fraction of cells able to take up glucose decreased from about 90% just after the treatment to zero during the first 24 h. The population that was still able to take up glucose during dark storage became increasingly less fluorescent, which indicates that the uptake rate of fluorescently labelled glucose decreased with time.

We never observed a regain of culturability in irradiated *S. typhimurium* or *S. flexneri* cells after 24 h or 48 h dark storage after irradiation with different light doses in laboratory or field experiments (data not shown).

Discussion

Susceptibility of *S. typhimurium* and *S. flexneri* to solar light. For the first time, cellular functions in *S. typhimurium* and *S. flexneri* during solar and UVA exposure were followed not only by plating but also with viability-staining and ATP measurements. The pattern of sequential loss of cellular functions during continuous artificial UVA exposure was similar in both tested organisms and corroborates our results reported for *E. coli* (Berney *et al.*, 2006a). The similarity in inactivation pattern suggests that the molecular mechanisms involved in the inactivation and killing of bacterial pathogens due to solar irradiation is similar or even identical in the three enteric bacteria. Therefore, *E. coli* can be considered a good model organism for such investigations. However, as suggested in an earlier study (Berney *et al.*, 2006b), both *S. flexneri* and *S. typhimurium* were found to be more resistant than *E. coli*.

Light dose reciprocity in SODIS. In experiments with *S. flexneri* and *S. typhimurium*, we observed two modes of exposure that caused deviation from the reciprocity law: firstly, exposure to either very high or very low irradiation intensity and secondly, split exposure by a pause in irradiation of up to 14 h. Interestingly, light dose reciprocity was valid for *E. coli* (Berney *et al.*, 2006b) and *S. typhimurium* (this study) within a much broader intensity range (50 - 700 W m^{-2}), whereas *S. flexneri* showed already distinct deviation when light intensities were two- to threefold higher than those occurring under natural conditions.

High irradiation intensity reduces light dose required for inactivation. Light dose reciprocity was observed in a majority of biological and medical applications, while so called "reciprocity law failures" were mostly observed in experiments conducted at either very low or very high radiant fluxes (Martin *et al.*, 2003). In the case of *S. flexneri* irradiated with UVA light we observed that the shoulder of the inactivation curve became less pronounced or was even eliminated with high irradiation intensities. The existence of a shoulder was interpreted earlier by either the presence of repair mechanisms that are able to slow down the light effect and, or damage to more than one single target within the cell (Harm, 1980; Sommer et al., 2001). If a shoulder is eliminated, as seen in our experiments for high intensity irradiation of S. flexneri and also S. typhimurium, repair mechanisms might be too slow or damaged themselves. Elimination of the shoulder results in a reduction of the light dose required for inactivation.

Split exposure reduces light dose required for inactivation. We found that S. typhimurium and S. flexneri are more susceptible to the same light dose when exposed to sunlight over two days (with a break over night) than with continuous artificial UVA light. This finding was reproducible in the laboratory with discontinuous irradiation on two consecutive days (data not shown). This is most likely because the bacterial cells are irreversibly damaged and consequently they continue to lose viability even when irradiation is stopped.

This is in line with the observation that exposure of cells from a human carcinoma cell line to short intervals of UVA light was more cytotoxic than continuous UVA irradiation (Merwald et al., 2005). However, if the time between exposures exceeded 2 hours, the cells were able to recover and, therefore, were less susceptible than with one single dose of UVA. Similar results were reported for Saccharomyces cervisiae and E. coli irradiated with UVC light (Dzidic et al., 1986; Harm, 1968; Salaj-Smic et al., 1985; Sommer et al., 1996) and human dermal fibroblasts irradiated with UVA light (tanning bed radiation) (Hoerter et al., 2008). Therefore, our data suggest that the repair mechanisms of S. typhimurium and S. flexneri were either already inactivated after 8 h of continuous irradiation, or that the lack of nutrients in the suspension did not allow the induction of an appropriate repair response. It remains to be determined whether or not the cells are able to recover when cell damage is less severe (shorter irradiation period) or when an appropriate amount of nutrients is available in the water.

Discontinuous UVC exposure of *E. coli* was shown to induce the SOS-response, which increases DNA repair activity (Dzidic *et al.*, 1986; Salaj-Smic *et al.*, 1985). However, UVA probably causes a more complex damage to the cells (several targets may be affected by UVA light, as compared to UVC, where predominantly DNA damages are observed) (Jagger, 1981). Hence, this should require a more complex repair-machinery. In an earlier microarray study we have shown that in *E. coli* UVA irradiation induces both DNA repair genes and genes involved in oxidative stress response upon irradiation with sublethal UVA light intensities (Berney *et al.*, 2006c).

Inactivation mechanism of SODIS. The data presented in this study strongly suggest that solar disinfection inactivates *S. typhimurium* and *S. flexneri* by inhibiting the respiratory chain. The ATP content per cell decreased rapidly in *E. coli* (Berney *et al.*, 2006a) and *S. flexneri* (this study) upon irradiation with sunlight, while in *S. typhimurium* an initial increase was observed before a sharp sustained decline. We propose that after an initial activation of energy metabolism, which is reflected in increased glucose uptake and ATP level, respiration stops and the remaining ATP is either consumed by various recovery processes (Kobayashi *et al.*, 2005), or, more likely, through the maintenance of the membrane potential via the F_1F_O-ATPase. Consistent with this proposal is the observed increase in glucose uptake activity, which provides ATP via substrate level phosphorylation to fuel the proton-pumping ATPase to maintain the membrane potential at a critical level even in the absence of a functioning electron transport chain. Since the medium used in this study (bottled mineral water) contains only low levels of assimilable organic carbon, the cells will eventually die from ATP exhaustion and loss of the membrane potential. In fact, it has been proposed earlier that components of the respiratory chain like menaquinones and dehydrogenases could be inactivated by UVA light (Jagger, 1981).

Repair and regrowth. In our experiments, injured bacterial cells irradiated with sunlight or UVA light were never observed to be able to regrow. This corroborates the work of other authors (Joyce *et al.*, 1996; Oates *et al.*, 2003; Reed, 1997a; Wegelin *et al.*, 1994).

The lack of regrowth in cells irradiated with polychromatic UV light in water disinfection processes seems to be the main advantage compared to monochromatic UVC light and therefore is of great interest to the water disinfection community (Kalisvaart, 2001; Kalisvaart, 2004; Oguma *et al.*, 2002; Zimmer-Thomas *et al.*, 2007).

It has been shown that non-cultivable cells of *S. typhimurium* produced by UVA irradiation do not retain infectivity for mice (Smith *et al.*, 2000). Our study now indicates that this is most likely due to an irreversible damage occurring during exposure to sunlight. Bacterial cells that are impaired in glucose uptake and oxidative phosphorylation may not regrow anymore, because uptake of nutrients and the maintenance of a membrane potential are regarded as prerequisites to survive and replicate.

Conclusion and implications for practice. Solar disinfection and artificial UVA light kill enteric bacteria most likely by inactivation of the respiratory chain and subsequent exhaustion of ATP. Our results show that even the resistant strain of *S. typhimurium*, which appeared to suffer only minor damage after half a day of sunlight, was actually damaged to an extent that regaining viability was not detected. In fact, our results suggest that it is even favourable to store the water over night before consumption, because injured cells will die from ATP exhaustion.

Acknowledgements

This project was financially supported by the Velux Foundation (project number 346). We thank Margarete Bucheli, Martin Wegelin, Regula Meierhofer, Silvio Canonica and Gregory M. Cook for valuable discussions.

3   The respiratory chain is the cells Achilles' heel during UVA inactivation in *Escherichia coli*.

Abstract

Solar disinfection (SODIS) is used as an effective and non-expensive tool to improve the microbiological drinking water quality in developing countries where no other means are available. Solar UVA light is the causative agent that inactivates bacteria during the treatment. The damage to bacterial membranes plays a crucial role in the inactivation process. In this study, the first targets on the way to cell death were found to be the respiratory chain and the F_1F_0 ATPase. The work suggests that protein damages are a likely cause for membrane dysfunction during UVA irradiation. Already slightly irradiated cells (after less than one hour of sunlight) are strongly affected in their ability to keep up essential parts of the energy metabolism.

This chapter has been published in *Microbiology* **156**, p2006-2015 by F. Bosshard, M. Bucheli, Y. Meur and T. Egli

Chapter 3

Introduction

The deleterious effect of sunlight on enteric bacteria is used to improve microbiological drinking water quality by solar disinfection (SODIS), a simple drinking water treatment method where the water is exposed to the sun in PET bottles for at least six hours. An epidemiological study during a cholera epidemic in Kenya showed a reduction of diarrhoea cases among SODIS users of 88% (Conroy et al., 2001). Although the effectiveness of SODIS against enteric pathogens is well documented, the underlying cellular inactivation mechanisms are not well understood. We know from earlier studies that a similar sequential inactivation pattern was found in different bacterial species like E. coli, S. typhimurium and S. flexneri (Berney et al., 2006a; Bosshard et al., 2009a) with total ATP contents and efflux pump activity lost early (> 500 kJ m^{-2} in E. coli), followed by a complete depolarisation of the cytoplasmatic membrane and loss of culturability (> 1500 kJ m^{-2}) and finally permeabilisation of the membrane (> 2000 kJ m^{-2}). Also for the highly resistant S. typhimurium, fluences corresponding to about half a day of sunlight (1500 kJ m^{-2}) were enough for further cellular die-off in the dark. Still it is not known which primary targets are severely enough damaged to kill the cell. It is also an open question when the cell crosses a „point of no return" where repair is not possible anymore, leading to further irreversible cell deterioration in the dark („dark inactivation"). Therefore, we are also addressing the question of when the damages accumulated during irradiation are irreversibly leading to cell death.

There is strong evidence that the aerobic energy metabolism of organisms is involved in initial inactivation and also die-off of irradiated cells. Energy metabolism is closely related to integrity of the membrane structure (intact lipid structure, fluidity of the membrane) and functionality of membrane proteins. Membranes allow the cells to generate ATP from an electrochemical potential via charge separation. Membrane proteins of the respiratory chain and the ATPase play an absolutely vital role in transforming the energy into ATP, a form that is used for many different cellular processes. But the electron transport to the terminal acceptor in the respiratory chain comprises also a danger for the cell since one-electron carriers are likely to generate reactive oxygen species (ROS) (Gianazza *et al.*, 2007). Many of the respiratory chain proteins rely on iron as a cofactor, which is a strong Fenton reactant and thus promotes the generation of hydroxyl radicals, clearly the most destructive of all ROS. They are causing damages in the near cell environment by lipid peroxidation (Niki *et al.*, 2005), but also to proteins where oxidation takes place as a consequence of ROS production (Bourdon & Blache, 2001). Lipid peroxidation induces a chain reaction by degrading polyunsaturated fatty acids into a variety of products, which can be very reactive and again are able to damage proteins (Cabiscol *et al.*, 2000). However, the indispensable prerequisite for a peroxidative chain reaction are polyunsaturated lipids, a feature that lacks in bacterial membranes (Imlay, 2009). Fenton reactions are accelerated by light (so called photo-Fenton reactions) and, therefore, radical damages speed up in the light (Chamberlain & Moss, 1987). The damaging effect of UVA light on membranes has been described previously (Moss & Smith, 1981) and the breakdown of energy metabolism within the cell was a likely explanation for cellular die-off (Berney *et al.*, 2006a; Bosshard *et al.*, 2009a). From the present state of knowledge, one can propose several reasons for breakdown of energy-generating systems within an aerobic cell: a) nutrient uptake is blocked, b) the generation of reducing equivalents by glycolysis or the TCA is blocked, c) respiration stops, d) F_1F_0 ATPase is inactivated, or d) the membrane is depolarised.

In this work, we are looking for the primary inactivation mechanisms during SODIS. Probable candidates are damages that are happening early and, at the same time, are vitally important for the cell. Therefore, we focussed on the energy-generating systems (respiration and ATP synthase activity) and the physiological response of the cells on an enzymatic level during the first hour(s) of sunlight irradiation.

Methods

Bacterial strains, growth media and cultivation conditions. In all experiments wild-type *E. coli* K12 MG1655 was used. For each experiment, a new tryptic soy agar (TSA; Biolife, Milano, Italy) plate was loop-streaked from cryo-cultures and incubated at 37 °C over night. For batch cultivation, LB broth (10 g tryptone, 5 g yeast extract, 10 g NaCl per litre) was filter sterilized (0.2 μm) and diluted to 1/3 of its original strength. Erlenmeyer flasks containing 20 ml of 1/3 LB were loop-inoculated with a single colony and incubated at 37 °C with vigorous shaking. When the cells reached exponential growth (OD_{546} between 0.1 and 0.2), the culture was diluted to an OD_{546} of 0.002 into 150 ml of prewarmed (37 °C) 1/3 LB in a 1000 ml Erlenmeyer flask and shaken for 18 hours. By that time the culture had entered stationary phase. Stationary phase was confirmed with five consecutive OD_{546} measurements within one hour.

UVA exposure. Cells were harvested by centrifugation (16000 g, 3min) and washed three times with filter sterilized bottled water (Evian, France). Depending on the experiment, cells were either diluted with filter sterilized bottled water to an OD_{546} of 0.02 and 0.2 (corresponding to ca. 2×10^7 cells ml^{-1} and 2×10^8 cells ml^{-1}, respectively) or kept at the original OD_{546} of 2.2 (ca. 2×10^9 cells ml^{-1}). Cell suspensions were then incubated for 1 - 2 hours at 37 °C to allow the cells to adapt to the bottled water. Then aliquots of 10 or 20 ml of cell suspension where exposed to UVA light in 30 ml quartz tubes. The tubes were placed in a carousel reactor (Wegelin *et al.*, 1994) equipped with a medium-pressure mercury lamp (TQ 718), which was operated at 500 W. The light emitted from the lamp passed through the glass jacket and through a filter solution before reaching the cells in the quartz tubes. The filter solution consisting of 12.75 g l^{-1} of sodium nitrate (cut-off at 320 nm) was used to obtain a UVA light spectrum comparable to solar light. The temperature of the filter solution was maintained at 37 °C during the experiments. The fluence rates at the position of the tubes were determined for each experiment with chemical actinometry using a volume of 10 or 20 ml

(Wegelin *et al.*, 1994). As unirradiated control, cells were kept in the dark at 37 °C for the maximum time of UVA exposure.

Cell fractionation: isolation of membranes vesicles and soluble fraction. Irradiated and non-irradiated samples with a sample volume of 20 ml and a concentration of $1 - 5 \times 10^9$ cells ml^{-1} were harvested and disrupted by passage through a pre-cooled French press at 15000 p.s.i. in 3 ml of membrane buffer (10 mM Mops, pH 7.2, 5 mM magnesium chloride hexahydride, 10% glycerol). Unbroken cells were then removed by centrifugation for 15 min at 16000 g. The remaining supernatant containing the bacterial membranes was spun again for 60 min at 135000 g at 4 °C and the resulting pellet was resuspended in 150 µl of membrane buffer. Aliquots of 50 µl were rapidly frozen in liquid N_2 and stored at - 80 °C (D'Alessandro *et al.*, 2008). The supernatant containing the soluble cell fraction was kept at -20 °C. Protein content of both membrane and soluble fraction was measured by the Bradford method (Bradford, 1976). Cells were irradiated at 5 different fluences in biologically independent triplicates.

Enzyme assays for soluble protein fraction. Enzyme assays were generally based on the spectrophotometric measurement (A_{340}) of the production or consumption of the cofactor NAD(P)H. Measurement time was 3 - 5 minutes , the temperature was kept at 22 °C and assay volumes were 3 ml containing either 10 or 100 µl of soluble protein fraction (protein conc. = 2 - 3 mg ml^{-1}). Background activity was determined in the absence of the substrate. Malat-dehydrogenase (EC 1.1.1.37): 100 mM phosphate buffer pH 7.3 was mixed with freshly prepared NADH (0.3 mM final concentration) and 10ul soluble protein fraction. The reaction was started with freshly prepared oxalacetic acid (0.2 mM final concentration) (Murphey *et al.*, 1969). Glucose-6-P dehydrogenase (EC 1.1.1.49): 55 mM TrisHCl pH 7.8 and 33 mM $MgCl_2$ were mixed with NADP (0.3 mM final concentration) and 100 µl soluble protein fraction. The reaction was started with glucose-6-P (3.33 mM final concentration) (John, 2002). Lactate dehydrogenase (EC 1.1.1.27): 100 mM phosphate buffer pH 7.5 was mixed with NADH (0.3 mM final concentration), pyruvic acid (0.2 % final concentration) and 100 µl soluble protein fraction (Tarmy & Kaplan, 1968). Glyceraldehyde-3-P dehydrogenase (EC 1.1.1.8): 50 mM phosphate buffer pH 8.5 was mixed with NAD+ (0.8 mM final concentration), Glyceraldehyde-3-P (0.8 mM final concentration), arsenate (5 mM final concentration) and 100 µl soluble protein extract (Allison & Kaplan, 1964). Glutathione-disulfide reductase (EC 1.6.4.2): 100 mM Hepes-KOH buffer pH 7.8 was mixed with EDTA (1 mM final concentration), NADHP (0.1 mM final concentration) and 100 µl cell extract. GSSG (1 mM final concentration) was added to start the reaction (Kunert et al., 1990); (Anderson et al., 1983).

Enzyme assays for membrane-associated proteins: respiratory activity by specific substrate-induced oxygen consumption. Enzyme assays for respiratory activity of membranes were performed using assays of 1ml total volumes in an oxygen electrode (Rank Brothers Ltd, Cambridge, England) in which the oxygen consumption was monitored with different substrates. Addressed were the enzymes NADH-ubiquinone oxidoreductase (EC 1.6.5.3), succinate oxidoreductase (EC 1.3.99.1) and L-lactate cytochrom c oxidoreductase (EC 1.1.2.3). The assay solution contained 50 - 150 µl membrane protein fraction in 50 mM phosphate buffer pH 7.5. Endogenous oxygen consumption was determined before addition of substrate. NADH (0.3 mM final) was added first, then succinate (1 mM final concentration) and lactate (1 mM final concentration). Before adding a new substrate, the measurement was followed until either a baseline was reached, or all oxygen was consumed and the assay solution was aerated until it was again saturated with oxygen.

ATPase activity. A commercial kit was used to evaluate ATPase activity of the membrane fraction (EnzCheck phosphate assay kit, molecular probes, Leiden, The Netherlands). The substrate 2-amino-6-mercapto-7-methylpurine riboside (MESG) was used as phosphate scavenger changing absorption from 330 to 360 nm, which allows quantification of inorganic phosphate released when ATP is hydrolysed by the F_1F_0 ATPase. The assay solution contained reaction buffer,

MESG substrate, purine nucleoside phosphorylase and membranes (10 µl ml^{-1}). The mixture was preincubated for 10 min to get rid of P_i background. When a steady baseline was reached, 4 µl of 10 mM ATP was added as a substrate for the ATPase and the reaction was followed at 360 nm. A P_i standard solution (500 µM) was diluted for a 6-point standard curve. Protein concentration of membrane fractions was 6 - 8 mg ml^{-1}. P_i background in the membrane fraction without addition of ATP was below detection limit.

Hydroperoxidase assay (HPI and HPII). The two *E. coli* hydroperoxidases or catalases are present either as a cytoplasmic membrane-associated enzyme (HPI, *katG* gene product) or as a cytosolic enzyme (HPII, *katE* gene product). Specific activities were measured in whole cell assays. The activity of the two enzymes can be separated by a heat inactivation treatment step because HPII is heat stable whereas HPI is heat labile (Visick & Clarke, 1997). Cells were washed and diluted in PBS to an OD_{546} of 1.25 and then exposed. To measure total catalase activity, prewarmed PBS was mixed with H_2O_2 (0.06 % final) and 0.5 ml cells (OD_{546} 1.25 in PBS). Absorption at 240nm was measured to follow the rate of decrease of H_2O_2. The assay volume was 3 ml and temperature was kept at 37 °C. To measure activity of HPI alone, HPII was inactivated prior to analysis at 55 °C for 15min.

ATP generation potential (ATP-GP). 900 µl cell suspension (10^7 cells ml^{-1}) were either incubated with 100 µl of prewarmed autoclaved full-strength LB or with 100 µl bottled water (as a control) at 37 °C for 1, 5, 10, 20, 40 min, respectively. The ability of the cells to produce ATP from the substrate was tested by measuring total ATP immediately after the incubation with the BacTiterGlo System (Promega, WI, USA) as described previously (Bosshard *et al.*, 2009a). The immediate response after 1min is very fast and gets linear only after the first five minutes of incubation ("steady state"). Therefore, ATP levels after 5, 10, 20 and 40min were considered for calculating the ATP production rate, or ATP generation potential (ATP-GP), and those data were used for the calculation of a linear regression curve ($\Delta ATP/\Delta t$). We decided against distinguishing between „free" and „cellular" ATP as fractions of „total" ATP (Hammes et al., 2008) since initial experiments showed that total ATP corresponded to intracellular ATP for our experimental settings and free ATP was found only as a neglectable fraction (< 5%).

Photostability of ATP. ATP was diluted to 10 nM in bottled water and immediately filter sterilized. Samples of 10 ml were irradiated and the ATP content of the samples was measured with the BacTiterGlo System immediately after sampling. An insignificant degradation of ATP was observed here when compared to ATP reduction in whole cells during UVA exposure. Maybe this is due to the presence of some small bacteria passing the 0.22 μm filter that may use up some ATP (Wang et al., 2007).

Plating. Aliquots were withdrawn at different time points during irradiation and during subsequent dark storage and diluted in decimal steps (10^{-1} to 10^{-5}) with sterile-filtered bottled water. Following dilution, 1 ml of test solution was withdrawn and mixed with 7 ml of liquid tryptic soy agar (TSA) (Biolife, Milano, Italy) at 40 °C (pour plate method). After 20 min, the solified agar was covered with another 4 ml of liquid TSA (40 °C). Plates were incubated for 48 hours at 37 °C until further analysis. Plate counts were determined with an automatic plate reader (acolyte, SYNBIOSIS, Cambridge, UK).

Results

ATP generation potential (ATP-GP) of cells when adding substrate after irradiation. To exclude the possibility of non-biological ATP degradation during irradiation experiments, degradation of pure ATP was measured in sterile-filtered bottled water without addition of bacterial cells. Only little degradation of pure ATP in bottled water was observed from exposure to UVA light compared to dark controls (<12%). Thus, photodegradation of ATP is not relevant in our experiments. In contrast to this, cellular ATP levels showed an initial transient increase to 110 % of the initial ATP content and then dropped massively with increasing UVA fluence (Fig. 3.1). This suggests that low light stess activates cells and induces ATP generation to enable repair of damages. Comparable to this, an increase of total ATP in cells exposed to low fluences was found before in *S. typhimurium* (Bosshard *et al.*, 2009a). The results shown here for *E. coli* further corroborate this hypothesis.

The ability to produce ATP is of vial importance for survival of cells during UVA irradiation. Therefore, we measured this ability in a whole cell assay where the cells are supplied with nutrients and the cellular ATP production rate was determined. This represents a whole cell ATP generation potential (ATP-GP). As nutrients, LB diluted to 10% of its original strength was provided. When added to non-irradiated cells, the substrate was converted over 40 min to a level of 180 x 10^{-7} ng ATP per cell, but this capacity was reduced to one third already for a fluence of 180 kJ m^{-2} and for fluences higher than this, it decreased further reaching a residual level at 900 kJ m^{-2} as low as 20 x 10^{-7} ng ATP per cell (Fig. 3.2). Moreover, when applying doses higher than 180 kJ m^{-2}, the production of ATP at 5, 10, 20 and 40 min after addition of substrate exhausted quickly, reached a final level and remained there constant. Further, culturability of the cells (at a cell density of $10^{\wedge 7}$ cells ml^{-1}) was reduced by about 90% with a fluence of 900 kJ m^{-2}, which corresponds well with earlier results (Berney *et al.*, 2006b).

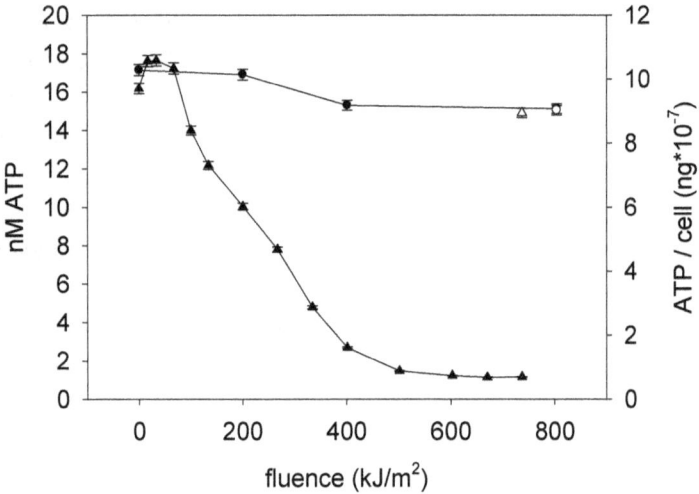

Fig. 3.1. Photostability of ATP in buffer (●, left axis) and of total cellular ATP in *Escherichia coli* during UVA irradiation (▲, right axis). Dark controls are displayed with empty symbols.

The ATP-GP of the cells was massively reduced to about 15% for an UVA exposure at 180 kJ m^{-2} and then continuously dropped down to 3% and below with fluences up to 900 kJ m^{-2} (Tab. 3.1). Without addition of substrate, the ATP regenerating activity was always very low in irradiated samples (< 2% of unirradiated control). For calculation of the ATP regenerating activity, the linear regression curve included only the data points between 5 and 40 min of incubation, since a fast initial ATP increase up to 20 x 10^{-7} ng ATP per cell was observed within the first 5 min of the assay. This increase was observed for fluences between 180 and 900 kJ m^{-2} (Fig. 3.2). When applying higher fluences (> 1200 kJ m^{-2}), this initial ATP boost decreased massively.

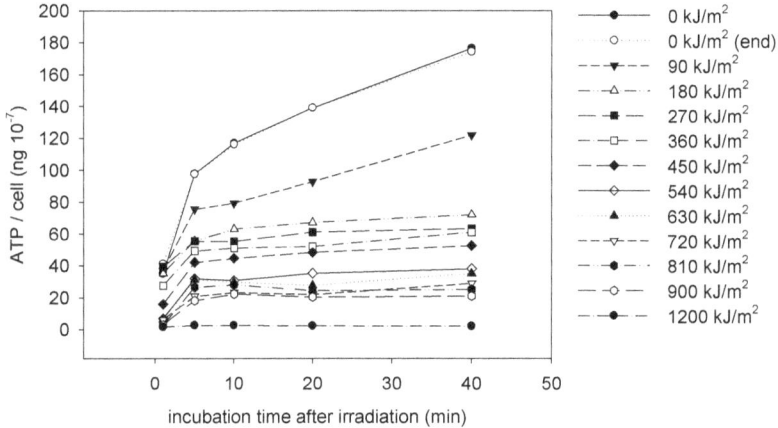

Fig. 3.2. ATP levels of cells exposed first to a defined UVA fluence and then immediately incubated in 10% LB medium final concentration for 1, 5, 10, 20 and 40 minutes. An unirradiated control (0 kJ m^{-2}) was measured in the beginning and in the end of the experimental time.

Enzymes of the inner membrane and hydroperoxidases. In order to achieve a high enough concentration of proteins for enzyme activity assays, 10^9 cells ml^{-1} were irradiated. Culturability of the cells decreased by about one log for a fluence of 900 kJ m^{-2} or by 7 logs for a fluence of 2000 kJ m^{-2} (Fig. 3.3a). Hence, both shoulder length and the degree of inactivation were similar to those achieved at lower cell densities, e.g. 10^7 cells ml^{-1}, indicating no significant effect of shading. All three assessed respiratory chain enzymes completely lost their activity very early in the irradiation process at 50 - 100 kJ m^{-2} (Fig. 3.3b), a fluence that corresponds to only 6 - 10 min of natural sunlight exposure. F_1F_0 ATPase activity decreased exponentially between 0 and 1000 kJ m^{-2} (Fig. 3.3c); it appears to be slightly less sensitive than the ATP-GP described above. Both hydroperoxidases dropped massively in activity with fluences up to 300 kJ m^{-2} (Fig. 3.4). Membrane-bound HPI lost its activity completely during the first 600 kJ m^{-2},

49

whereas the cytosolic HPII maintained some residual activity around 5 - 10% of the unirradiated control.

Tab. 3.1. ATP generation potential (ATP-GP) of UVA irradiated *E. coli* cells with addition of substrate (LB 10% final concentration) within 40 min after the addition of substrate. The immediate response after 1min is very fast and gets linear only after the first five minutes of incubation ("steady state"). Therefore, only ATP levels after 5, 10, 20 and 40min were considered for calculations on the ATP generation potential and those data were used for the calculation of a linear regression curve indicated as $\Delta ATP/\Delta t$.

Fluence (kJ m^{-2})	with nutrients (10% LB)			control without nutrients		
	$\Delta ATP/\Delta t$ (nM min^{-1})	$\Delta ATP/(\Delta t*OD_{546})$ (nM min^{-1})	%	$\Delta ATP/\Delta t$ (nM min^{-1})	$\Delta ATP/(\Delta t*OD_{546})$ (nM min^{-1})	%
0	3.59	17.958	100.00	0.21	1.027	5.72
0 (end)	3.68	18.4015	102.47	0.15	0.75	4.18
90	2.30	11.5075	64.08	0.03	0.15	0.84
180	0.42	2.08	11.58	0.03	0.1315	0.73
270	0.70	3.5105	19.55	0.05	0.2315	1.29
360	0.42	2.08	11.58	0.06	0.2975	1.66
450	0.57	2.838	15.80	0.02	0.115	0.64
540	0.50	2.504	13.94	0.01	0.0325	0.18
630	0.34	1.7225	9.59	0.02	0.083	0.46
720	0.21	1.067	5.94	0.01	0.0365	0.20
810	0.11	0.559	3.11	0.01	0.0365	0.20
900	0.06	0.3155	1.76	0.01	0.0645	0.36
1120	0.00	0.0105	0.06	0.01	0.0505	0.28

Enzymes of the cytoplasm. Cytosolic enzymes generally maintained their activity longer than enzymes in other compartments (Fig. 3.3c). Malat dehydrogenase activity was almost unaffected by irradiation up to a fluence of 1600 kJ m^{-2}, lactate dehydrogenase and glucose-6-P dehydrogenase activity were only reduced to about 50%. Exeptions were glutathione reductase and glyceraldehyde-3-P dehydrogenase that lost their activity completely (<1%). This is consistent with a previous study on glyceraldehyde-3-P dehydrogenase, where a reduction to 10% of the initial activity was found with an applied UVA fluence of 1000 kJ m^{-2} (Voss *et al.*, 2007).

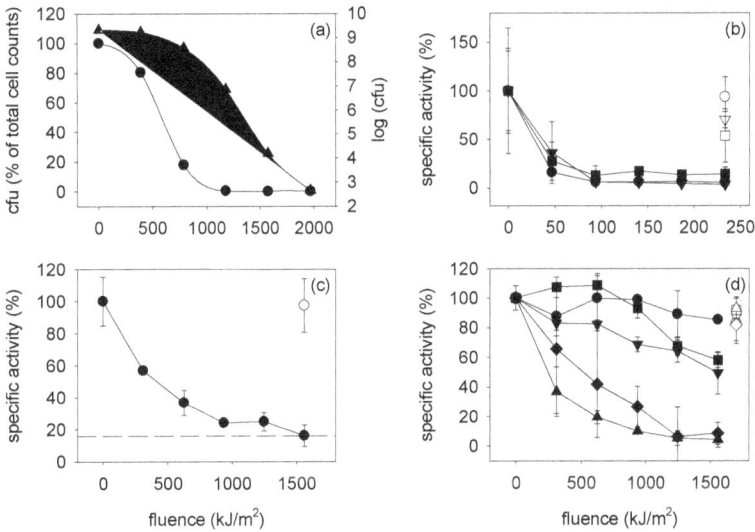

Fig. 3.3. Culturability and specific enzyme activities in UVA irradiated *E. coli* cells. Cell suspensions irradiated had a cell density of 10^9 cells ml^{-1} and activities are depicted as % of activity found in untreated cells. (a) Culturability shown in % of total counts (•) and on a log (c.f.u ml^{-1}) scale (▲). (b) activity of respiratory chain enzymes determined by specific substrate-induced oxygen consumption in the membrane fraction (ΔmgO$_2$ (min^{-1} mg protein^{-1}) ml^{-1}): NADH-oxidase (•), succinate oxidase (▼), lactate oxidase (★) (100% correspond to 4.75, 4.68 and 0.63 mgO$_2$ min^{-1} (mg protein^{-1}) ml^{-1}, respectively. (c): F$_1$F$_0$-ATPase activity in the membrane fraction. 100% correspond to 47.52 µM P$_i$ min^{-1} (mg protein)$^{-1}$ ml^{-1}. The dashed line shows background activity of the membrane fraction with the F$_1$F$_0$-ATPase specific inhibitor DCCD. (d): activity of enzymes in the cytoplasm: malate DH (•), lactate DH (■), glucose-6P DH (▼), glutathione-disulfide reductase (♦), glyceraldehyde-3P DH (▲). 100% correspond to ΔAbs340 min^{-1} mmol^{-1} cm^{-1} (mg protein)$^{-1}$ ml^{-1} of 0.406, 0.045, 0.375, 0.0198, and 0.0177, respectively. Empty symbols represent dark controls.

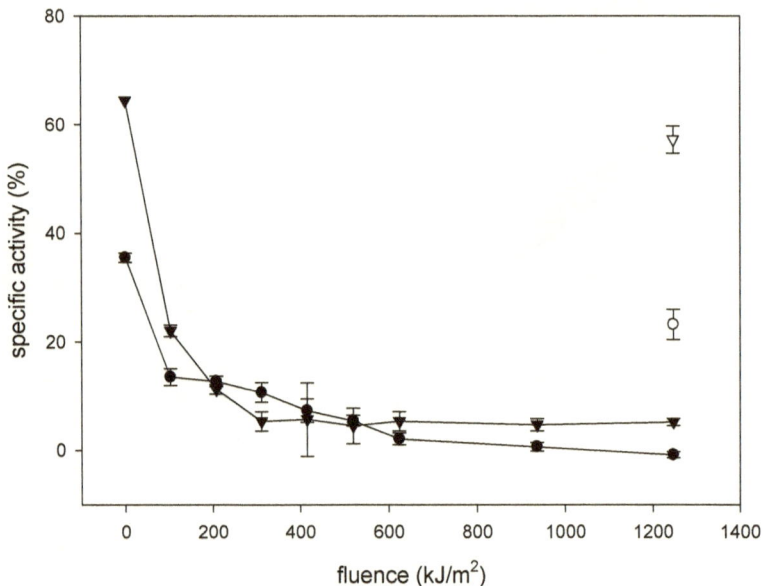

Fig. 3.4. Specific hydroperoxidases activities $(\Delta A_{240}/(min*\Delta OD_{546})$; 100% corresponds to 0.0173 $min^{-1})$: HPI (•), HPII (▼). Empty symbols represent dark controls.

Dark inactivation after UVA irradiation. An earlier study revealed that dark storage after half a day of sunlight exposure (1500 kJ m^{-2}) was progressively killing not only *E. coli*, but also *S. typhimurium* cells, a bacterium that is less susceptible to sunlight than other enteropathogens (Bosshard *et al.*, 2009a). To study this process of dark inactivation in more detail and obtain a better time resolution, we used *E. coli* as a model organism. The cells were irradiated in the lower fluence range to elucidate whether or not a threshold fluence exists at which the cell is damaged only reversibly and able to recover and repair the damages accumulated during irradiation. The inactivation of *E. coli* cells in artificial UVA light followed an inactivation curve with an initial shoulder up to 800 kJ m^{-2}, where no effect on cell culturability was observed, followed by a phase of log-linear inactivation with about a 1.5-log reduction at 950 kJ m^{-2}. Cells irradiated with UVA up to 150 kJ m^{-2} and subsequently stored in the dark for up to 48h showed no effect and behaved as the unirradiated control (after 50, 100, 150 kJ m^{-2}). At higher doses, a massive drop in culturability of one log (after 300, 480 kJ m^{-2}), two to three logs (after 950, 1500 kJ m^{-2}) or close to the detection limit (after 1900 kJ m^{-2}) was observed (Fig. 3.5). The results demonstrate that after a fluence of about 300 kJ m^{-2} no repair or recovery did occur.

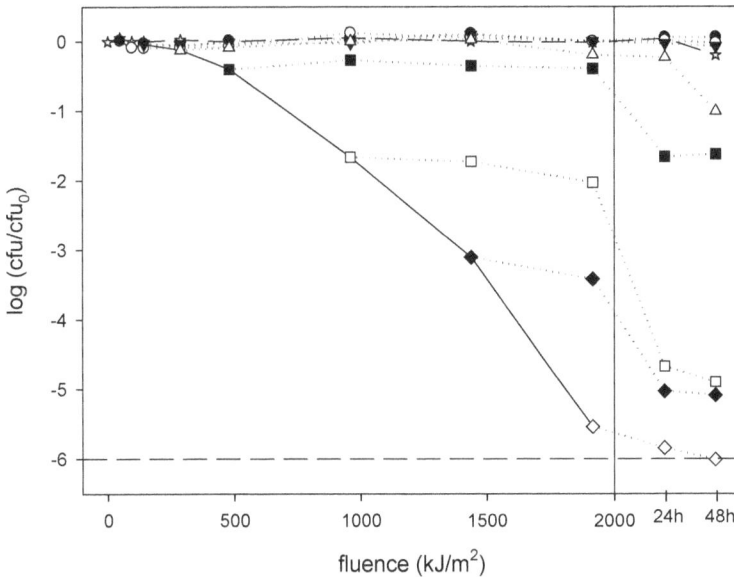

Fig. 3.5. Irradiation and subsequent dark storage of UVA irradiated *E. coli* cells. Cells were exposed at a cell density of 10^7 cells ml^{-1} to fluences of 50 (●), 100 (○), 150 (▼), 300 (△), 480 (■), 950 (□), 1500 (◆) and 1900 kJ m^{-2} (◊) during the first 7 hours of the experiment and stored in the dark thereafter (indicated by the dotted line). A non-irradiated control is also shown (☆). At t_{24} and t_{48} after the start of the experiment, samples were plated again to determine the degree of survival (box at the right, for 24 and 48 hours of storing in the dark). The experiment was repeated three times, representative results are shown. The dashed line represents the detection limit.

Discussion

Changes in cellular energy metabolism during UVA irradiation. The temporal inactivation pattern of different cellular functions during SODIS is illustrated in a conceptual graph (Fig. 3.6). The breakdown of respiration precedes all other tested crucial cellular functions such as the massive decrease in catalase activity, the depletion of the cellular ATP pool, the loss of ATPase activity, and culturability. Membrane depolarisation and permeabilisation were observed only later in the inactivation process as reported already earlier (Berney *et al.*, 2006a). Special emphasis on the loss of essential cellular processes therefore focus on energy metabolism, a subject that will be discussed in more detail below.

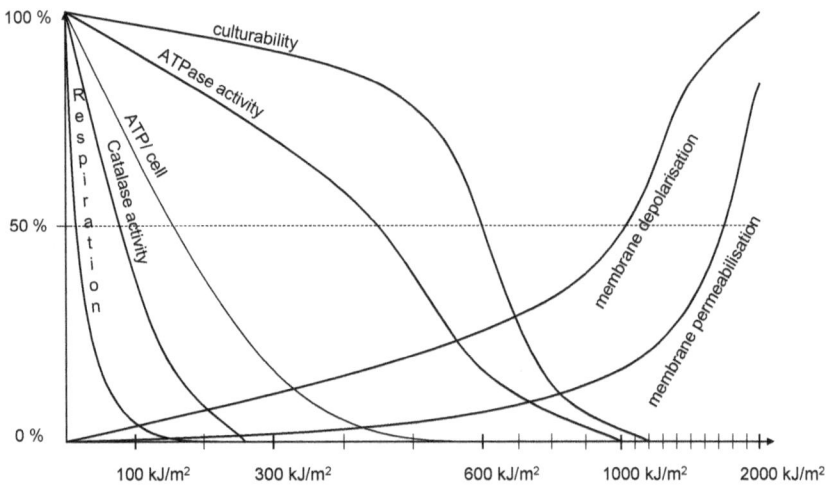

Fig. 3.6. Conceptual graph showing the temporal pattern of inactivation by UVA light in stationary *Escherichia coli* cells. Catalase activity represents combined activity of HPI and HPII. Values for membrane depolarisation and permeabilisation are taken from (Berney *et al.*, 2006a). One tick corresponds to a fluence of 100 kJ/m².

A cell's ability to maintain its ATP level is essential for coping with any stress situation where it needs readily available energy for defence and repair of damages. A decrease in the ATP generation potential (ATP-GP), therefore, is fatal for a cell under stress. For this reason, we assessed the cell's immediate response to restore cellular energy (ATP) in the presence of nutrients. To our knowledge, this is a new approach to evaluate cellular viability. Even at an exposure as low as 180 kJ m^{-2}, corresponding to less than half an hour of sunlight, the cells were massively affected in their ability to immediately produce ATP (Fig. 3.2, Tab. 3.1). However, if incubated on agar plates, more than 90% of the cells were still able to repair their damages and to produce a colony after exposure to 180 kJ m^{-2} (Fig. 3.3a). In contrast, the ability to produce ATP was virtually inexistent when cells irradiated with fluences of about 1200 kJ m^{-2} (Fig. 3.2), a fluence at which culturablity on agar plates drops to < 1% (Fig. 3.3a). These observations might be explained by a breakdown of energy metabolism during UVA irradiation in three phases. First, the cells were probably still able to transport substrates and link this to ATP production. This corresponds to the phase where an initial ATP increase at fluences below 180 kJ m^{-2} was observed (Fig. 3.1). Second, at fluences exceeding 180 kJ m^{-2}, the cells still had a residual capacity to produce ATP and recharge their cellular ATP pool with a bottom level that was in the range of 20 x 10^{-7} ng ATP per cell (Fig. 3.2). In this phase, the processes that use ATP were probably still slower than those that resynthesize it. ATP was probably used in this phase for restoration of the membrane potential, which was maintained up to fluences of 1000 kJ m^{-2} at levels >50% of its original value (Fig. 3.6) and maybe also for repair processes. Given sufficient time, with incubation up to 48h on agar plates, most of the cells were still able to recover from this state of damage and and grew to visible colonies. In the third phase at fluences higher than 1200 kJ m^{-2}, the cells lost the capacity to restore the cellular ATP pool; now ATPase activity dropped massively and also the membrane started to become leaky, as measured in an earlier study by live/dead-staining (Berney *et al.*, 2006a). Culturability decreased exponentially at fluences higher than 900 kJ m^{-2} after an initial shoulder (Berney *et al.*, 2006d). This suggests that the cell tries to maintain as long as possible the membrane potential, which is important for many of substrate transport processes across the membrane, via ATPase.

The fact that cells lose their ability to regenerate ATP has to be interpreted with caution because several ATP consuming and producing metabolic processes influence the pool. The ATP pool is subject to turnover and we did measure neither the turnover rate nor the single processes that use or generate ATP separately. Processes involved in ATP consumption are uptake of nutrients by membrane transport systems, repair, restoration of membrane potential; processes involved in ATP generation are oxidative phosphorylation and substrate level phosphorylation. Contributions of individual ATP consuming processes are difficult to measure and therefore were not investigated here. To get a rough idea of the underlying processes, we focussed on investigating individual components and enzymes involved in generation of ATP, like oxidative phosphorylation via F_1F_0 ATPase, glycolysis (represented by glucose-6-P dehydrogenase and glyceraldehyde-3-P dehydrogenase), the TCA (represented by succinate dehydrogenase and malate dehydrogenase) and respiration (represented by NADH oxidase, succinate oxidase and lactate oxidase activity of the cytoplasmatic membrane). Furthermore, nutrient uptake is an essential prerequisite that fuels all of these processes with electron donors.

Respiratory membrane enzymes were targeted very early during irradiation. The functioning of the respiratory chain is very essential for aerobic cells since, together with F_1F_0 ATPase, it is the most efficient way to produce a membrane potential and ATP. NADH dehydrogenase and succinate dehydrogenase contain FeS-centres; cytochromes contain haems and FeS-centres. In *E. coli*, all dehydrogenases tested translocate electrons to a central ubiquinone/ubiquinol pool in the membrane, which then gets oxidized by either cyt bo_3 (a haem-copper protein that has a low oxygen affinity and is more expressed at high O_2 saturation levels and, therefore, is more relevant for our cells), or cyt bd (containing 2 b-type haemes as acceptors and a d-type haem as oxygen reduction site with high oxygen affinity and expressed when oxygen is limiting) (Nicholls & Ferguson, 1992; Unden & Dünnwald, 2008). The fact that all dehydrogenases are inactivated at similar fluences suggests that not each of these enzymes is inactivated independently, but that the electron transport chain is interrupted. The electron transport chain in mitochondria has been widely investigated in connection with oxidative stress. It was found that in the mitochondrial electron transport chain, the flavin mononucleotide group of complex I and not the ubiquinone of complex III is the primary physiological relevant site of ROS generation (Yuanbin *et al.*, 2002). The electron transport chain of *E. coli* has not been examined with this focus so far, but since electron transport chains are evolutionarily quite conserved, flavin enzymes are suspected as primary target of ROS damage during UVA irradiation in the respiratory chain of *E. coli*.

Direct measurement of ATPase activity showed that this function is maintained for fluences about ten times higher than respiration. In yeast, near-UV was proposed to hamper ATP synthase activity by photodecomposition of ergosterol leading to a more rigid membrane (Arami *et al.*, 1997a; Arami *et al.*, 1997b). Bacteria have a class of molecules in their membranes with a similar structure and function, called hopanes (Neidhardt *et al.*, 1990), which might be sensitive to UV light too. Moreover, it is believed that also lipid peroxidation decreases membrane fluidity and is able to disrupt membrane-bound proteins. Lipid peroxidation causes a damaging chain reaction that amplifies oxidative stress by forming more radicals when degrading polyunsaturated fatty acids into a variety of products, e.g. aldehydes. They can be very reactive and damage then again other molecules such as proteins (Cabiscol *et al.*, 2000). However, the indispensable prerequisite for a peroxidative chain reaction are polyunsaturated lipids, a component that lacks in bacterial membranes (Imlay, 2009).

The ability to take up nutrients is essential for the cell's survival. The ability to take up a fluorescent glucose analogue was tested in an earlier study which showed that this function remained active even up to higher fluences than the ATPase (Berney *et al.*, 2006a; Bosshard *et al.*, 2009a); the reason for this is probably that this analogue is transported by the PTS system that is fed via substrate-level phosphorylation. Different studies showed that near-UV affects the uptake of amino acids into the cell (Ascenzi & Jagger, 1979; Robb *et al.*, 1978; Sharma & Jagger, 1981). Preliminary results from Biolog-AN plates (data not shown) indicated that not only amino acids, but also the ablity to take up and oxidize many different substrates was reduced early during irradiation. The ability for reduction of substates in the assay got lost simultaneously for all C-sources, again indicating that the final step of the respiratory chain is inactivated at this time rather than single transport systems within the membrane (Bochner, 2009).

The cytoplasmic metabolic enzymes of the glycolysis and most enzymes of the TCA cycle) are part of the soluble fraction of the cell. These were active for considerably higher fluences than membrane enzymes, some were even virtually unaffected by irradiation. Generally, the susceptibility of the soluble enzymes tested varied over a wide range. In contrast to membrane enzmyes, none of them had metal cofactors. Lactate and malate dehydrogenases also exhibited a high resistance to ozone treatment in an earlier study (Komanapalli *et al.*, 1997), which corresponds to the high UVA resistance in our experiments. In contrast, glyceraldehyde-3-P dehydrogenase lost most of its activity during ozone treatment, as it did during UVA treatment here.

Cellular defence against ROS during UVA irradiation. Hydroperoxidases and glutathione-disulfide reductase are part of the ROS defence of the cell. Glutathione-disulfide reductase activity stayed basically unaffected during UVA irradiation. This enzyme was found to be highly resistant to UVA and ozone treatment before (Hoerter *et al.*, 2005a; Komanapalli *et al.*, 1997). In contrast, both types of hydroperoxidases (HPI and HPII) contain a haem-cofactor (Claiborne & Fridovich, 1979; Claiborne *et al.*, 1979). These iron centres are probable ROS generators and this may explain the fast inactivation of the two enzymes in UVA light. In addition, our results show that the membrane-associated and heat-labile HPI lost its activity even earlier than the cytoplasmic HPII. Furthermore, the higher heat resistance of HPII was argued to be due to stabilizing elements in the structure and the containment of haem d instead of haem b (Switala & Loewen, 2002). These factors may be explanations for higher resistance of HPII to UVA light as well.

Inactivation in the dark after irradiation. The breakdown of enzymatic processes observed here must reach a point where the cell would become unable to repair the damages and as a consequence of energy limitation, the cell would be lethally damaged. We were interested in the point of "no return", where irreversible cell damage was reached. Therefore, we irradiated cells with relatively low UVA fluences and subsequently stored them in the dark. We assumed that irreversibly damaged cells would continue dying when stored, while reversibly damaged cells would be able to repair their damages as soon as they were plated and obtained nutrients. Indeed a threshold fluence for irreversible damages was observed: at fluences below 300 kJ m^{-2} the cells were able to recover from damages. After this, even if the inactivation process was still in the „shoulder" phase (lasting up to 900 kJ m^{-2}), where repair should theoretically be possible (Harm, 1980), they progressively lost culturability in the dark when irradiation was stopped. Even when supplied with nutrients, the cells were unable to refill their ATP pool since the regeneration capacity itself was damaged, as shown in a previous section. This implies for practice that cells do not recover from UVA light damages, even after short irradiation times or lower fluences due to overcast conditions or increased water turbidity.

ROS mediated damage to membrane proteins probably causes cell death during SODIS. Our results corroborate the inactivation mechanism proposed previously (Bosshard *et al.*, 2009a) which comprised an early inhibition of the respiratory chain and the ability to maintain the membrane potential by reverse use of the ATPase. Indeed, ATPase activity dropped at the same time as the membrane potential decreased, as measured by $DiBAC_4(3)$-staining (Berney *et al.*, 2006a). Therefore, we can conclude with some certainty that membrane proteins are the first targets during SODIS. Already with a fluence of only 50 kJ m^{-2} respiratory chain enzymes in the membranes were inactivated and for fluences > 300 kJ m^{-2} the cell damages definitely became irreversible. In the cell, the respiratory chain is responsible for about 90% of the $OH^{\bullet}$ radical production under normal growth conditions (Gonzalez-Flecha & Demple, 1995). $OH^{\bullet}$ radicals are short-lived only and react in the near surrounding where produced (Latch & McNeill, 2006). This corroborates that membrane proteins (respiratory enzymes, F_1F_0 ATPase) and lipids are especially in danger of oxidation by this species, an assumption also made before (Choksi *et al.*, 2008). One could argue that with an interrupted respiratory chain the electron flow and thereby also ROS stress would immediately be stopped. However, when looking at, e.g., the still increasing protein oxidation levels with fluences >50 kJ m^{-2} (Bosshard *et al.*, 2009c), it seems clear that some form of oxidative stress continues after inhibition of respiration. The continuing ROS production with a non-functional respiratory chain is that the still active electron transport chain continues to produce ROS by dumping electrons into the surrounding instead of a controlled delivery to the terminal acceptor. Another possibility is that in the presence of a non-functional respiratory chain, reducing agents are accumulated within the cell and drive ROS production again by metal-catalyzed oxidation reactions. These are known to be fuelled within the cell by the reduced NAD(P)H in the presence of oxygen and iron. The reducing equivalents thereby are used to regenerate the metal (Levine, 2002). A third possibility is that an increased ratio in TCA cycle to respiration activity boosts the production of ROS, a common mode of action observed with many antibiotics (Kohanski *et al.*, 2007). DNA damage was observed during near-UV irradiation, too (Jiang *et al.*, 2009), but only after irradiation at relatively high fluences (> 800 kJ m^{-2}). Therefore, protein damage must be considered as an early and very essential target during SODIS cell inactivation.

Acknowledgements

This project was financially supported by the Velux Foundation (project Nr. 346). We thank Frédéric Gabriel for practical help with the oxygen electrode and Frederik Hammes for valuable discussions about ATP experiments.

4 Protein oxidation and aggregation in UVA-irradiated *Escherichia coli* cells as signs of accelerated cellular senescence

Abstract

Solar disinfection (SODIS) is a simple drinking water treatment method that improves microbiological water quality where other means are unavailable. It makes use of the deleterious effect of solar irradiation on pathogenic microbes and viruses. A positive impact on health has been documented in several epidemiological studies. However, the molecular mechanisms damaging cells during this simple treatment are not yet fully understood. Here we show that protein damage is crucial in the process of inactivation by sunlight. Protein damages in UVA irradiated *Escherichia coli* cells have been evaluated by an immunoblot method for carbonylated proteins and an aggregation assay based on semi-quantitative proteomics. A wide spectrum of structural and enzymatic proteins within the cell is affected by carbonylation and aggregation. Vital cellular functions like the transcription and translation apparatus, transport systems, amino acid synthesis and degradation, respiration, ATP synthesis, glycolysis, the TCA cycle, chaperone functions and catalase are targeted by UVA irradiation. The protein damage pattern caused by SODIS strongly resembles the pattern caused by reactive oxygen stress. Hence, sunlight probably accelerates cellular senescence and leads to the inactivation and finally death of UVA irradiated cells.

This chapter has been published in *Environmental Microbiology* **12**, p2931-2945 by F. Bosshard, K. Riedel, T. Schneider, C. Geiser, M. Bucheli and T. Egli

Chapter 4

Introduction

Sunlight is an important environmental factor that has influenced life since it appeared on earth. In many surface ecosystems microorganisms are transiently affected by sunlight, a prominent example are the surface waters of the sea (Jeffrey *et al.*, 2005). Bacterial enteric pathogens have the guts as their primary habitat and are therefore usually more sensitive to solar radiation than environmental strains that had the chance to adapt during evolution. The deleterious effect of solar radiation on enteric bacteria is used to improve microbiological drinking water quality by solar disinfection (SODIS), a simple drinking water treatment method (Wegelin *et al.*, 1994). A positive impact on health has been documented in several epidemiological studies, e.g. during a cholera epidemie in Kenia, where a reduction of diarrhea cases among SODIS users of 88% was observed (Conroy *et al.*, 2001). SODIS was recently added to the WHO list of recommend drinking water treatment methods. Although the effectiveness of SODIS against enteric bacterial pathogens is well documented, the underlying cellular inactivation mechanisms are not yet well understood.

Protein oxidation is known to be a key factor in cellular aging in eukaryotes (Grune *et al.*, 2004) and was recently also found to be important in bacteria (Nyström, 2006). The tertiary structure of oxidized proteins is thermodynamically instable and therefore, oxidized proteins tend to expose hydrophobic amino acids to the outside, with the consequences of agglutination and crosslinking (Chiti, 2006; Grune *et al.*, 2004; Squier, 2001). The accumulation of protein aggregates is associated with age-related diseases and senescence in many different organisms (Mazzulli *et al.*, 2006), also in bacteria, where aggregation was suggested to lead to cell death (Maisonneuve *et al.*, 2008a). Reasons for the accumulation of protein aggregates in a cell can result from either an increased level of damaged proteins or due to a less efficient protein degradation and repair. In eukaryotic cells the proteasome is in charge of protein turnover and removal of oxidized proteins, and it might even itself be a target of oxidative stress (Friguet,

2006). Damaged proteins in aggregates are not easily accessible for proteases. Indeed, it was suggested that protein fragmentation and crosslinking might make proteins resistant to proteolytic digestion (Gianazza *et al.*, 2007). In bacterial cells, carbonylated proteins are predominantely found in protein aggregates where they seem to escape degradation (Maisonneuve *et al.*, 2008b).

Very early loss of energy generating systems within the cytoplasmatic membrane (respiration and ATPase activity) appear to be a primary cause for cellular die-off during sunlight irradiation (Bosshard *et al.*, 2009b). This loss of enzymatic functions suggests that protein damages at the cytoplasmatic membrane are involved in cell inactivation during SODIS. The underlying cause for protein damage most probably is oxidative stress. For example, it was reported that reactive oxygen species (ROS) originate predominantely from „leakage" in the respiratory chain (Yuanbin *et al.*, 2002) and therefore, is believed that respiratory chain enzymes are also the first targets of oxygen stress (Choksi *et al.*, 2008). This is also supported by the observation that adaptation to UVA light clearly induced many genes of the oxidative defence system (Berney *et al.*, 2007b). There is much evidence that proteins are a very important target of oxidative damage within cells. Whereas lipid peroxidation and DNA modification for a long time was the most investigated radical-mediated process, more recent studies indicate also proteins as important targets of oxidative damage with severe concequences for cell functioning (Bourdon & Blache, 2001). Surprisingly, protein damage seems to be crucial in highly resistant ionizing radiation-resistant bacteria such as *Deinococcus radiodurans* where the degree of resistance appears to be determined by the level of oxidative protein damage caused during irradiation. The authors are convinced that protein and not DNA is the principle target of the biological action in sensitive bacteria and that extreme resistance is based on protein protection (Daly *et al.*, 2007). One protection mechanism in *D. radiodurans* is to reduce cytosolic iron ions up to three times and, instead, to increase manganese ions up to 300-times. By this, the cells are capable to reduce iron-dependent Fenton reactions causing protein oxidation and therefore to protect enzymatic functions. Another more common line of defence against oxidative damage found in many cell types, e.g. in *Escherichia coli*, are the enzymes catalase, superoxide dismutase and peroxidases (Fredrickson *et al.*, 2008; Gross, 2007).

Oxidized proteins are believed to severly impair cell viability (Berlett & Stadtman, 1997). The most widespread method to assess the level of oxidatively damaged proteins within a cell is by measuring the level of newly introduced carbonylation sites in proteins (Requena *et al.*, 2003). Detection by a western blot assay is used after blotting of one- or two-dimensional gels (Levine, 2002). Several authors have used this assay method successfully in microorganisms, after UVA treatment (Hoerter *et al.*, 2005a; Hoerter *et al.*, 2005b), with oxidative stress (Cabiscol *et al.*, 2000; Tamarit *et al.*, 1998a) and with senecence and aging (Dukan & Nystrom, 1998). Carbonylation is a basic damage leading to covalent crosslinking of proteins and the formation of protein aggregates by several mechanisms (Stadtman & Levine, 2003). These covalently crosslinked protein aggregates are stable in SDS PAGE because they withstand the reducing conditions (by DTT or mercaptoethanol) and are not affected by the detergent (SDS), a fact that was already utilized early to examine oxidation effects of hydroxyl radicals on several purified enzymes. An increase in protein size was interpreted as a sign of protein aggregation, a decrease as protein fragmentation (Davies, 1987; Davies & Delsignore, 1987). Other authors used advanced protein identification methods instead of gel electrophoresis to detect oxidatively crosslinked and fractionated proteins (Mirzaei & Regnier, 2007).

In this study, a proteome analysis of *E. coli* cells irradiated with UVA light in a reproducible laboratory setting is presented. It gives an indication of which proteins are damaged during SODIS and an insight into the underlying mechanisms leading to cell death during exposure to sunlight is obtained.

Methods

Bacterial strains, growth media and cultivation methods. In all experiments wild-type *E. coli* K12 MG1655 was used from cryo-cultures and for each experiment loop-streaked onto a new TSA agar plate. For bach cultivation, LB broth (10g tryptone, 5g yeast extract, 10g NaCl, per litre) was filter-sterilized with Millipore syringe filters (Millex GP, 0.2 μm) and diluted to 1/3 of its original strength. Erlenmeyer flasks containing 20ml of diluted LB were inoculated with a single colony and incubated at 37°C with vigorous shaking until the cells reached exponential growth (OD_{546} between 0.1 and 0.2). Then the culture was then diluted to an OD_{546} of 0.002 into 150ml of prewarmed diluted LB in a 1000ml Erlenmeyer flask and shaken over night for 18 hours until stationary phase was reached. Stationary phase was confirmed with five consecutive OD_{546} measurements.

UVA exposure. Cells were harvested by centrifugation at 16000xg for 15 min and washed three times with filter-sterilized still mineral water. The pellet was suspended in filter-sterilized mineral water to reach an OD_{546} of 2.2 (1 - 5x 10^9 cells ml^{-1}) in order to obtain a high enough protein concentration. Cell suspensions were incubated for 1 hour at 37°C to allow the cells to adapt to the mineral water. Aliquots of 20ml of cell suspension were exposed to UVA light in 30 ml quartz glass tubes. The tubes were placed in a carousel reactor (Wegelin *et al.*, 1994) equipped with a medium-pressure mercury lamp (TQ 718), which was operated with 500 W. The light emitted from the lamp passed through the glass jacket and through a filter solution before reaching the cells in the quartz glass tubes. The filter solution, consisting of 12.75 g l^{-1} of sodium nitrate (cut-off at 320nm), was used to obtain a UVA light spectrum comparable to solar light. The temperature of the filter solution was maintained at 37°C during the experiments. The fluence rates at the position oft the tubes were determined during every experiment by using actinometry (Wegelin *et al.*, 1994). For each experiment, a part of the bacterial suspension was kept at 37°C as a dark control.

Whole cell extract. Triplicate samples (irradiated at 1000 kJ m^{-2} and non-irradiated) with a sample volume of 20ml and a concentration of 1 - 5x 10^9 cells ml^{-1} were harvested by centrifugation at 16000x g for 15 min, resuspended in 3ml of lysis buffer (10mM Hepes pH7, 0.1% CHAPS, 1x Protease inhibitor (complete, Roche, Mannheim, Germany)) and disrupted by one passage through a pre-cooled French press at 15000 p.s.i.. Cell disruption efficiency with one passage through the french press was always > 97% in non-irradiated and irradiated samples as evaluated by microscopy and flow cytometric total cell counts. Unbroken cells were then removed by centrifugation for 15min at 16000xg and glycerol was added to a final concentration of 5%. Samples were rapidly frozen in liquid N$_2$ and stored at -20°C. Protein concentration was measured by the Bradford method (Bradford, 1976).

1D-SDS PAGE combined with LC-MS/MS. Whole cell extracts were separated by SDS PAGE (Laemmli, 1970) using 12% polyacrylamide gels and stained by Coomassie (Gallagher, 2006). Protein lanes were cut in twelve gel slices (A - M), which contained approximately the same amount of proteins. Proteins were immediately subjected to in-gel tryptic digestion (Shevchenko *et al.*, 1996). The resulting peptide mixtures were analyzed on a hybrid LTQ-Orbitrap mass spectrometer (ThermoFischer Scientific, Bremen, Germany) interfaced with a nanoelectrospray ion source. Chromatographic separation of peptides was achieved on an Eksigent nano LC system (Eksigent Technologies, Dublin, CA, USA), equipped with a 11 cm fused silica emitter, 75 μm inner diameter (BGB Analytik, Böckten, Switzerland), packed in-house with a Magic C18 AQ 3 μm resin (Michrom BioResources, Auburn, CA, USA). Peptides were loaded from a cooled (4°C) Spark Holland auto sampler and separated using an acetonitrile/water solvent system containing 0.1% formic acid at a flow rate of 200 nL/min with a linear gradient from 3 to 35 % acetonitrile in 60 min. Up to 6 data-dependent MS/MS spectra were acquired in the linear ion trap for each Fourier-transform (FT)-MS spectral acquisition range. The latter was acquired at 60,000 full-width half-maximum (FWHM) nominal resolution settings with an overall cycle time of approximately 1 second. Charge state screening was employed to select for ions with two charges and rejecting ions in single-charge state. The automatic gain control (AGC) was set at $5e^5$ for ion injection control and at $1e^4$ for full FT-MS and linear ion trap MS/MS. The instrument was calibrated externally according to the manufacturer's instructions. All samples were acquired using internal lock mass calibration on m/z 429.088735 and 445.120025.

Database searches, data validation and protein quantification. The Mascot Search Engine (version no. 2.2.04) was used for protein database searches against the Swissprot reference database. MS/MS ion searches were performed with the following settings: (i) trypsin was chosen as protein-digesting enzyme and up to two missed cleavage sites were tolerated, (ii) carbamidomethylation of cystein was chosen as fixed modification, and (iii) oxidation of methionine and formation of pyro-glutamic acid from glutamine and glutamic acid were chosen as variable modifications. Searches were performed with a parent ion mass tolerance of 5 ppm and a fragment ion mass tolerance of 0.8 Da. Scaffold (version 2.05.01, Proteome Software Inc., Portland, OR, USA) was used to validate and quantify MS/MS-based peptide and protein identifications. Peptide identifications were accepted if they could be established at greater than 95.0% probability as specified by the Peptide Prophet algorithm (Keller *et al.*, 2002). Protein identifications were accepted if they could be established at greater than 99.0% probability and contained at least one identified peptide. Protein probabilities were assigned by the Protein Prophet algorithm (Nesvizhskii *et al.*, 2003). Proteins that contained similar peptides and could not be differentiated based on MS/MS analysis alone were grouped to satisfy the principles of parsimony. Semi-quantitative analyses of protein abundances were performed on the basis of the number of unique peptides that were assigned to a protein by the Scaffold software (= spectral counts). The spectral counts were then blotted for each individual protein against the molecular sizes of the gel pieces (A - M). For the 200 most abundant proteins, enough unique peptides were found to assess the original molecular weight of the protein and a possible shift after UVA irradiation due to aggregation or fragmentation. Carbonylation sites were found by filtering results for amino acid mass shifts of +15.999 for the conversion of proline to glutamic semialdehyde, -43.071 for the conversion of arginine to glutamic semialdehyde and -1.031 for the conversion of lysine to aminoadipic semialdehyde (Requena et al., 2003).

Immunoblot against carbonylation sites. Western blotting was performed to detect carbonylated proteins using the Oxyblot™ Protein Oxidation Kit (Millipore, Chemicon, Zug, Switzerland) according to the manual. The samples containing 7.5 or 15 ug of protein were subjected to 2,4-dinitrophenylhydrazine (DNPH) to derivatise carbonylation sites to 2.4-dinitrophenylhydrazone. Solutions for negative control samples lacked DNPH. After 15min exposure to the derivatising reagent a neutralisation solution was added until a stable change in colour from yellow to red was observed. The DNPH-derivatized protein samples were separated by 12.5% or 8% SDS PAGE and blotted onto nitrocellulose membranes (Bio-Rad, Reinach, Switzerland). The membranes were blocked with 1% of bovine serum albumin (BSA) and PBS-T and incubated with primary antibodies against DNP in 1% BSA in PBS-T for 1 h at room temperature. This was followed with a horseradish-peroxidase antibody conjugate directed against the primary antibody for 1 h at room temperature. The chemiluminescent visualisation was carried out as described earlier (Shacter et al., 1994). After the exposure to the film the membrane was incubated with chromogenic BM Blue POD substrate (Roche, Rotkreuz, Switzerland). Like this it was possible to assemble the film on the membrane and draw the prestained MW marker onto the film. This step was necessary because the carbonylated MW marker provided by Chemicon was not consistently detectable and the obtained pattern could not always be related to the MWs given in the protocol. Computerized densitometry was performed using a GS-800 calibrated Densitometer with the 1D Analysis software Quantity one® (both from Bio-Rad).

Centrifugable aggregates by „mild centrifugation". As reported recently, (Maisonneuve *et al.*, 2008a; Maisonneuve *et al.*, 2008c), we separated cellular protein aggregates from whole cell extracts obtained by French press with relatively mild centrifugation (16000xg) and measured the protein content of the supernatant at different time points after onset of centrifugation. Protein loss in the supernatant was interpreted as protein loss due to sedimentation of insoluble proteins in the form of large aggregates.

Dissolved organic carbon (DOC) measurements. UVA irradiation with 2×10^8 cells ml^{-1} was done like described above. Cells were removed by filtration through pre-washed 0.22 µm filters. DOC was then measured on a SHIMADZU TOC-5050A analyzer with a highly sensitive catalyst as non-purgeable organic carbon (NPOC).

2D-gel electrophoresis. 1ml samples with a cell concentration of $1 - 5 \times 10^9$ cells ml^{-1} were taken from the quartz glass tubes or from dark controls into Eppendorf tubes placed on ice, centrifuged at 4°C and the pellet was resuspended in 50 µl lysis buffer I (stock solution composition: 50 µl 0.5 M Tris-HCl pH 6.8, 80 µl 15 % SDS, 20 µl glycerol, 40 µl beta-mercaptoeptanol and 270 µl nanopure H2O). Cells were incubated for 4 min at 95°C, followed by 1 h of shaking at 37°C. After incubation, 50 ul lysis buffer II (9.5 M Urea, 0.04 % Nonidet P40 and 5 % β mercaptoeptanol) was added, the sample was fastfrozen (liquid nitrogen) and stored at -80°C until gel electrophoresis was performed. To each sample 50 U of benzoase (Merck KGaA, Darmstadt, Germany) were added and nucleic acids digestion was performed for 1 h at 37°C. Protein concentration was measured using Bio-Rad Protein Assay Kit, 200 µg proteins were added to rehydration buffer (accordingly to the manufacturer) adjusted to an end volume of 330 µl and used to rehydrate ReadyStrip pI 3-10 (Bio-Rad, Basel, Switzerland). Active rehydatation was performed with IEF-Cell (Bio-Rad) for 12 h at 50 V, followed by isoelectric focusing at 9000 V until 60000 V*h was reached. Equilibration and transfer to the second dimension were performed according to the manufacturer

(Bio-Rad). SDS PAGE was performed on acrylamide gels (12.5%) using PROTEAN II MULTI-CELL (Bio-Rad). Gel images were scanned with the GS-800 Calibrated Densitometer (Bio-Rad) and the obtained images were analyzed with the PDQuest Basic software version 8.0 (Bio-Rad) (Franchini, 2006).

Results

Protein damage during SODIS was characterized in this paper with the help of several different methods. Carbonylation damages are detected already with very low fluences by a sensitive immunoblot method. These oxidative damages are prerequisites for protein aggregation, which we detected at higher fluences with gel electrophoresis. Tandem mass spectrometry in combination with a semi-quantitative spectral count analysis helped to identify proteins that tend to aggregate or fragment during SODIS. Furthermore, a massive loss of protein spots was observed on 2D gels, which brought us to test the possibility of leaking of the cells during SODIS. DOC analysis revealed that only a very small percentage of cellular C was leaking out during irradiation. Interestingly, in raw extracts of irradiated cells, mild centrifugation was able to separate a big part of the cellular proteins as insoluble protein aggregates.

Qualitative aggregation effect on whole cell proteome assessed by gel electrophoresis. One surprising effect of UVA irradiation was that cellular proteins became unable to properly separate with traditional gel electrophoresis methods after the treatment. With the aim to identify damaged proteins, we first naively intended to compare unirradiated and irradiated cell samples on high resolution 2D-gels. Surprisingly, a massive loss of protein spots (70%), smearing and shifting of protein spots in irradiated samples as compared to the unirradiated control was observed (Fig. 4.1A, B). This effect was reproducible. The shifting of protein spots of a similar molecular weight along the isoelectric focusing dimension was described before as „protein stuttering" due to mistranslation (Ballesteros *et al.*, 2001), in our case the alteration in amino acid side chains probably leads to similar effects. Furthermore, on one-dimensional gels, a smearing of protein bands was observed starting at an irradiation dose of around 1000 kJ m^{-2} (Fig. 4.1B, lane 6) and this effect became even more pronounced with increasing irradiation doses. Gels were prepared with a reducing agent (DTT) and SDS as a detergent and, therefore, only covalently

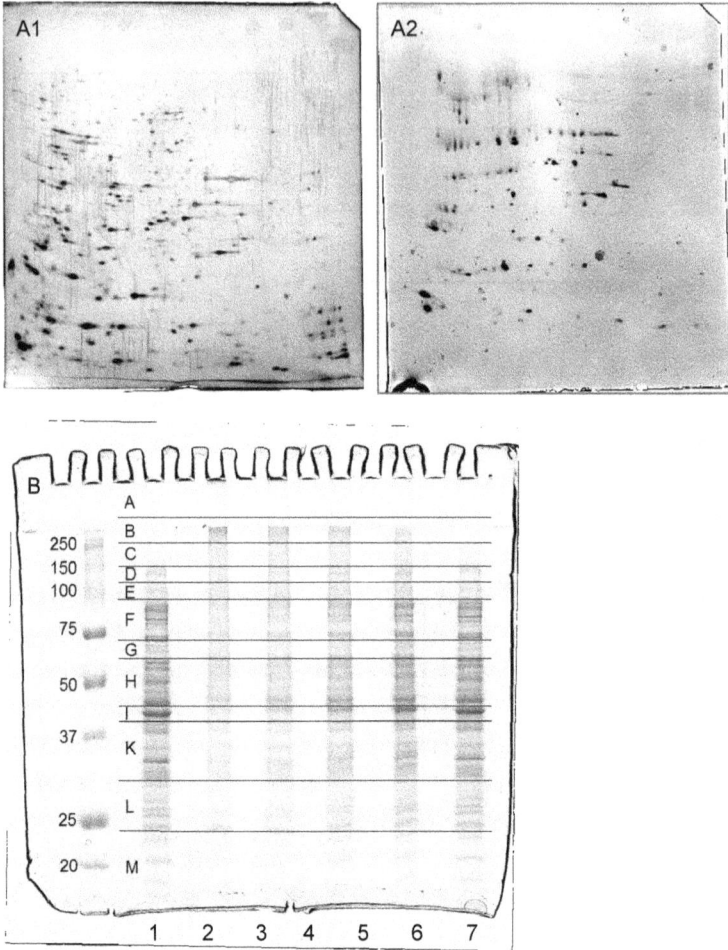

Fig. 4.1. One- and two-dimensional gelelectrophoreses of UVA-irradiated *E. coli* samples and unirradiated dark controls. A: 2D-gelelectrophoresis of unirradiated control (A1) and a sample irradiated with a fluence of 2200 kJ m^{-2} (A2). B: 1D-SDS-PAGE with protein standard ladder (lane 1; 250, 150, 100, 75, 50, 37, 25, 20 kDa) unirradiated control (lane 2) and irradiated samples with fluences of 2750 kJ m^{-2} (lane 3), 2200 kJ m^{-2} (lane 4), 1650 kJ m^{-2} (lane 5), 1100 kJ m^{-2} (lane 6), 550 kJ m^{-2} (lane 7). Experiments were done at least three times, representative results are shown. In A, 150 ug of cellular protein was loaded per gel. In B, 15 ug of cellular protein was loaded onto gels per lane.

crosslinked aggregates would stick together whereas S-S bridges or hydrophobic interactions would be solubilized (Davies, 1987; Lund *et al.*, 2008). High irradiation doses make this effect even more pronounced (Fig. 4.1B, lane 3), with a loss of cellular protein and the formation of a new „band" at the transition from the stacking to the separating gel.

DOC measurements of irradiated whole cells. Since many protein spots were missing on 2D gels and an obvious decrease in proteins resolved on 1D gels of irradiated cells was observed also, DOC was measured in supernatants of irradiated cells to be able to exclude the possibility of cellular protein loss during irradiation. The cell concentration used in these experiments (2×10^8 cells ml^{-1}) corresponds to a dry weight of about 56 ug ml^{-1}, from which only about 1 µg ml^{-1} cellular C was lost with increasing UVA irradiation (Fig. 4.2). As a cell contains about 50% C (28 ug ml^{-1}), this corresponds to a loss of cellular C into the supernatant of about 2.8%. Even if this C exclusively originated from cellular protein (and not other cell components like DNA, lipids or polysaccharides), this loss would only account for about 5.5% of the whole cellular protein content, since proteins on average consist of 46% C (Neidhardt *et al.*, 1990).

Quantitative aggregation effect on whole cell protein content. Since proteins seemed to stay within the cells according to DOC results, we wondered whether it would be possible to quantify the protein aggregation effect that obviously lead to the „protein loss" observed on gels. Therefore, cells were completely distroyed and the protein content of the whole cell extract was assessed after various times of mild centrifugation. During mild centrifugation, only larger insoluble particles settle whereas fully soluble proteins stay in solution and, therefore, account for the protein content of the whole cell extract. When applying irradiation doses of 1000 kJ m^{-2} and higher, larger protein aggregates were formed that one could obviously spin down, reaching as much as 80% of the total protein (Fig. 4.3).

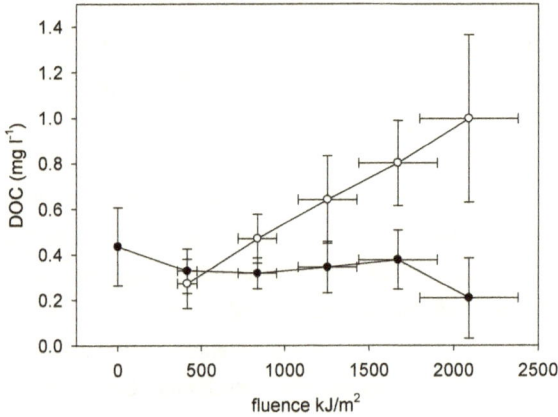

Fig. 4.2. Loss of DOC into surrounding medium during irradiation of 10^8 cells/ml with (•) non-irradiated controls and (○) irradiated samples.

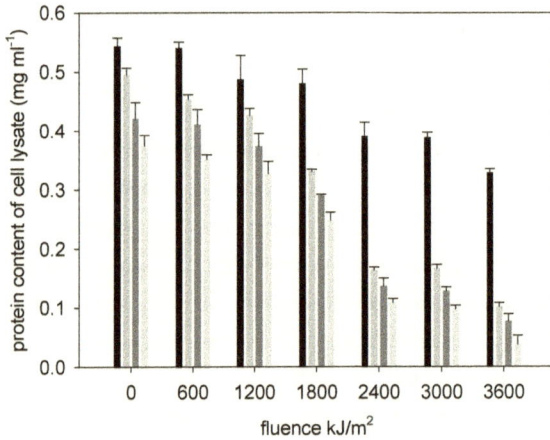

Fig. 4.3. Loss of whole cell protein by mild centrifugation: Protein concentration in the whole cell lysate (black bar) and in the supernatants after 20, 40, 80min of mild centrifugation at g-force of 16000 (middle grey, dark grey, light grey bar, respectively). A dark control (0 kJ m^{-2}) was measured at the beginning and at the end of the experiment.

Interestingly, also the total protein content in the non-centrifuged whole cell lysate was decreasing to about 60% during irradiation, while the cell disruption efficiency still stayed between 97 - 99%. This might be due to amino acid modifications that alter the staining properties of proteins. DOC measurements showed that cells only leak a small amount of their total C content during UVA irradiation and, therefore, one must conclude that the proteins are still in the whole cell extract but not stainable anymore.

Identification of aggregation target proteins on 1D-SDS PAGE. As an alternative to 2D-gels, we identified proteins affected by UVA light with the help of 1D-gels. Samples irradiated with a dose of about 1000 kJ m^{-2} and unirradiated controls were separated by 1D-SDS PAGE. In order to reduce sample complexity, the gel lanes were sliced in twelve gel pieces (A - M) always containing proteins in the same molecular range (Fig. 4.1B). The proteins within these gel slices were then identified by tandem mass spectrometry. Interestingly, the number of proteins identified in the high molecular range, especially of gel pieces of stacking gel and upper part of separating gel (according to molecular weights of >250 kDa) was increasing with higher irradiation doses (Fig. 4.4A), reaching a maximum of 45% of all identified proteins present in these high molecular weight gel fractions. The same effect was visible when looking at the whole molecular weight range (for the gel slices A (>250 kDa) to M (<25 kDa)), where an increase in the number of identified proteins is observed in the higher molecular weight ranges, while the number decreases within the low molecular weight ranges (Fig. 4.4B).

Aggregation on single proteins evaluated by semi-quantitative spectral counts analysis. The spectral count is the number of unique peptides found for a particular protein. Assuming that the number of unique peptides is proportional to the amount of a protein in the original sample, spectral counts can be used as a semi-quantitative measure for the amount of the protein in the original sample. Here, the spectral counts were blotted for each individual protein against the corresponding molecular weight of the gel slices (A - M) to screen for possible mass shifts due to aggregation damages in the irradiated samples compared to the non-irradiated controls. For the 200 most abundant proteins, enough unique peptides were found to assess the original molecular weight of the protein and a possible mass shift after UVA irradiation due to aggregation or fragmentation. One example for a protein with a mass shift is the alpha-subunit of the F_1F_0 ATPase (Fig. 4.5). 71 of the 200 most abundant proteins of known function were aggregating reproducibly in three independent experiments. Their functions and the corresponding shifts are shown in Tab. 4.1. Many different cellular functions are affected in this way (Fig. 4.6).

A

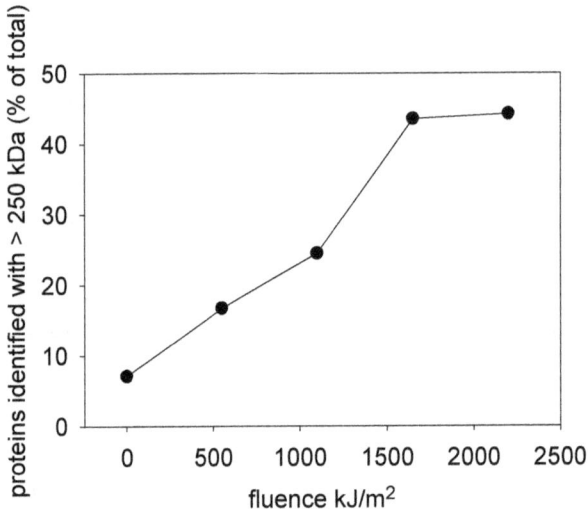

B

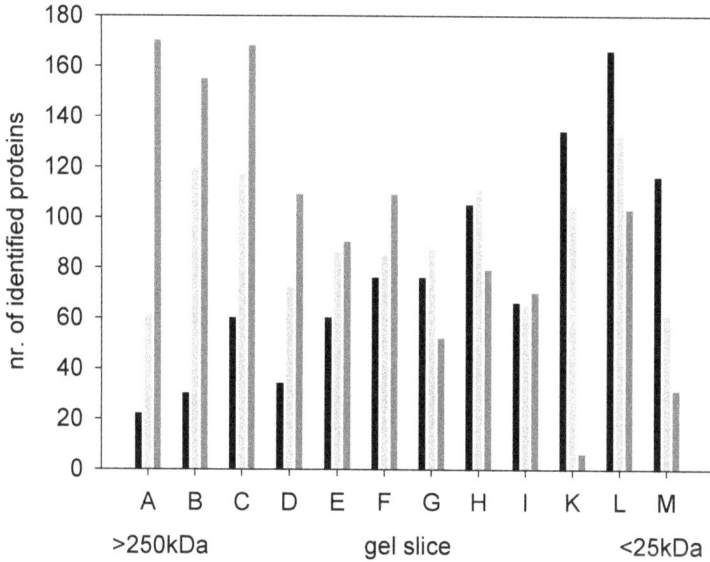

Fig. 4.4. The gel cuts of the higher molecular weight range increase in the number of identified proteins when cells have been irradiated. A: Number of identified proteins in fractions of a molecular size of > 250 kDa, with increasing fluence. B: Fractions over a molecular weight range of > 250 kDa to <25 kDa. Shown are unirradiated samples (black) and irradiated samples at fluences of 1100 kJ m^{-2} (light grey) and 2200 kJ m^{-2} (dark grey). These results correspond to lane 1, 3 and 5 in Fig. 4.1B, respectively).

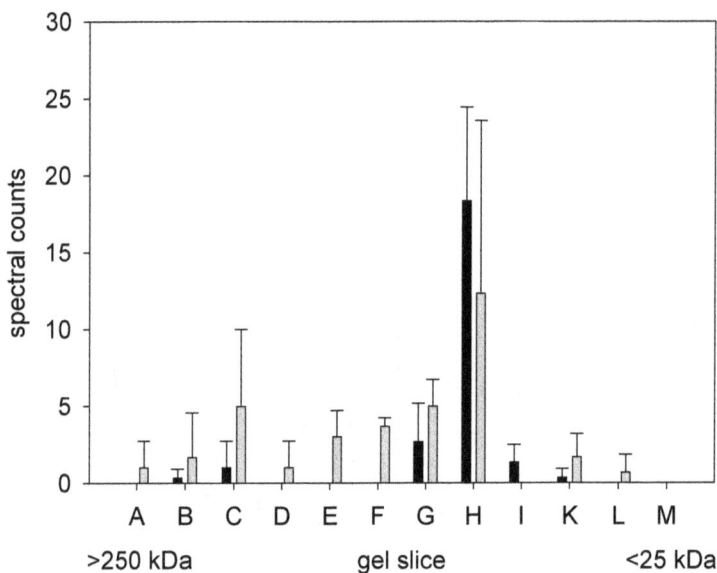

Fig. 4.5. The alpha-subunit of the F_1F_0-ATPase as an example for a protein that was analyzed to aggregate during irradiation. Black bars represent distribution of the protein in the unirradiated sample, grey bars the irradiated sample. Error bars indicate standard deviations in spectral counts of a triplicate measurement.

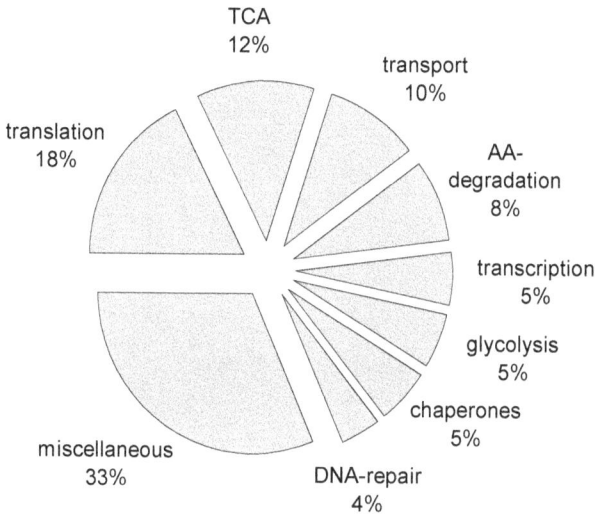

Fig. 4.6. Cellular functions of the 71 proteins found with a mass shift (corresponding to 100%) out of the 200 most abundant *E. coli* proteins.

Tab. 4.1. List of proteins found with a mass shift. These proteins were reproducibly aggregating in three independent experiments. Corresponding gel lanes are shown in Fig. 4.1A, lane 2 for control and lane 6 for irradiated samples (1100 kJ m^{-2}).

protein name	function	MW (kDa)	found in gel slices (control)	found in gel slices (irradiated)
RNA polymerase subunit beta	transcription /transcription regulation	155	CDE	ABCDE
ATP-dependent protease Lon		87	F	BCDEF
Transcription termination factor Rho		47	I	BCDEFHI
RNA polymerase sigma factor RpoD		70	F	BCEF
Elongation factor Tu	translation/ protein	43	HIKLM	ABCDEFGHI KLM
Elongation factor G	biosynthesis	78	EFG	ABCDEFG

Ribonuclease E		118	CDE	ABCDE
Translation initiation factor IF-2		97	EF	BCDEF
30S ribosomal protein S1		61	FG	ABCDEFG
Alanyl-tRNA synthetase		96	EF	ABCDEF
Glycyl-tRNA synthetase beta subunit		77	F	BCDEF
Threonyl-tRNA synthetase		74	F	BCDF
30S ribosomal protein S7		20	M	BDELM
Phenylalanyl-tRNA synthetase beta chain		87	F	BCEF
50S ribosomal protein L5		20	M	AM
50S ribosomal protein L6		19	M	ABCGM
Magnesium-transporting ATPase	transport/ chemotaxis	99	CDEFGH	ABCDEFGH
Phosphoenolpyruvate-protein phosphotransferase		64	G	ABCDEFG
Preprotein translocase subunit secA		102	EF	BCDEF
Acriflavine resistance protein B		114	EF	BCDE
Acriflavine resistance protein A precursor		42	I	BCDI
PTS system mannitol-specific EIICBA component		68	H	BCEFGH
PTS system glucose-specific EIICB component		51	IK	CDEFI
Ribonucleoside-diphosphate reductase 1 subunit alpha	replication	86	EF	BCDEF
Carbamoyl-phosphate synthase large chain	AA synthesis	118	CDE	ABCDE
Glutamine synthetase		52	GH	ABCDEFGH
Cysteine desulfurase		45	IK	ABCDI
Tryptophanase	AA degradation	53	GHIKL	ABCDEFGHIKL
4-aminobutyrate aminotransferase		46	HI	ABCDEFGHIK
Bifunctional protein putA		144	DE	ABCDE
Glycine dehydrogenase		104	EF	ABCDEF
Aminopeptidase N		99	G	ABCEF
Aspartate ammonia-lyase		52	H	ABCDEGH

Polyribonucleotide nucleotidyltransferase	RNA degradation	77	F	BCDEF
3-oxoacyl-[acyl-carrier-protein] synthase 1	lipid metabolism	43	I	ABCHI
Inosine-5'-monophosphate dehydrogenase	purin/ pyrimidine metabolism	52	H	BCDEGH
Bifunctional purine biosynthesis protein PurH		57	GH	BCDEGH
NADH-quinone oxidoreductase subunit G	respiration	100	EF	BCDEF
ATPase subunit alpha	ATP synthesis	55	GHI	ABCDEFGH KL
Enolase	glycolysis	46	HIKL	ABCDEGHI KL
Fructose-bisphosphate aldolase class 2		39	BK	BEFGKL
Pyruvate kinase II		51	H	BCDGH
2-oxoglutarate dehydrogenase E1 component	TCA	105	EF	ABCDEF
Pyruvate dehydrogenase E1 component		100	EF	ABCDEF
Aconitate hydratase 2		94	EF	ABCDEF
Succinate dehydrogenase flavoprotein subunit		64	FG	ABCDEFG
Dihydrolipoyl dehydrogenase		51	H	ABCDEFGH
Acetyl-coenzyme A synthetase		72	FG	ABCDEFG
Citrate synthase		48	HIKL	ABCDEFGHI KL
Dihydrolipoyllysine-residue acetyltransferase component of pyruvate dehydrogenase complex		66	FG	ABCDEF
Succinyl-CoA synthetase beta chain		41	IKL	ABCDEGIKL

Dihydrolipoyllysine-residue succinyltransferase component of 2-oxoglutarate dehydrogenase complex		44	H	ABCDEFGH
Formate acetyltransferase 1	fermentation	85	FG	BCDEFG
Aldehyde-alcohol dehydrogenase	degradation of CH	96	EFG	ABCDEFG
glucose-1-phosphatase		46	HI	CDEFGHI
Isocitrate lyase	glyoxylate cycle	48	HIKLM	ABCDEFGHI KLM
Transketolase 1	pentose phosphate pathway	72	F	BCFG
Transketolase 2		73	F	BCF
Phosphoenolpyruvate carboxykinase [ATP]	gluconeogenese	60	GHL	BCDEFGHL
Phosphoenolpyruvate synthase		87	G	BCEF
NADP-dependent malic enzyme	c4 metabolism	82	F	BCDEF
Chaperone protein DnaK	chaperone	69	FG	ABCDEFG
Chaperone protein ClpB		96	F	ABCDEFG
Trigger factor		48	H	ABCDEFGH
Chaperone protein HtpG		71	FG	BCDFG
Peroxidase/catalase HPI	catalase	80	F	BCDEF
DNA gyrase subunit A	DNA repair	97	EF	BCDEF
DNA gyrase subunit B		90	F	BEF
DNA protection during starvation protein		19	M	ACDEFGIKL M
Osmotically-inducible protein Y precursor	osmotic stress	21	M	ADEFGIKLM
Periplasmic oligopeptide-binding protein precursor	periplasm	61	GH	ABCDEFGH
Outer membrane protein A precursor		37	KL	ABCDEFGHI KL

csiD predicted protein	unknown	37	K	ABCDEFK
glutamate decarboxylase		53	H	BCDEFGH
alpha				

Carbonylation damage to proteins. Oxidative carbonylation damage to proteins was observed very early during irradiation, and the pattern is not changing thereafter (Fig. 4.7). Overall, carbonylation damage increases 3.5 fold when compared to dark controls in samples irradiated with 645 kJ m^{-2} and 7.5 fold when irradiated up to 2500 kJ m^{-2} (Tab. 4.2 and 4.3, respectively). The bands on the Western blot seem rather specific up to an irradiation dose of 950 kJ m^{-2}, which corresponds to the dose where the cells start dying off when plated on TSA agar. Two reproducible bands around 78 kDa and 55 kDa are found with the carbonylation immunoblot method already with an irradiation dose of 215 kJ m^{-2} and a third band of about 35 kDa is appearing with irradiation doses of 645 kJ m^{-2} and more (Fig. 4.7A). In the high molecular weight range (>170 kDa) a pronounced accumulation of carbonylated sites is detected (Fig. 4.7A) and with irradiation doses higher than 950 kJ m^{-2} a general „smearing" similar to the one on SDS PAGE is visible on the blots as well (Fig. 4.7B, C), both indicating a high amount of carbonylation sites in aggregating proteins. The identification of carbonylated proteins after UVA irradiation turned out to be challenging. Two different approaches were taken. First, samples were run on 2D gels with the intention to do immunoblotting, as reported before (Dukan & Nystrom, 1999). However, proteins of irradiated cells shifted so strongly that the picture was not comparable to the original pattern anymore (Fig. 4.7B). Secondly, carbonylated proteins were labelled with biotin-hydrozine with the intention to specifically enrich them by avidin for further analysis on SDS PAGE and tandem-mass spectrometry (Mirzaei & Regnier, 2005). However, this affinity enrichment was not as selective as expected, be it due to non-specific background biotinylation within our samples, be it' due to protein aggregation making the specific enrichment difficult. Nevertheless, we tried to figure out probable candidates for early carbonylation targets with two criteria. First, they needed to have a molecular size in the range of the positively stained oxyblot bands and secondly, they needed to show mass shifts making them suspicious to contain newly introduced carbonylation sites. Putative carbonylation candidates are namely elongation factor G (EFG 78 kDa), 60kDa chaperonine (GroL 57 kDa), ATPase subunit alpha (ATPA 55 kDa), ATPase subunit beta (ATPB 50 kDa), Lysyl-tRNA synthetase (SYK2 58 kDa), glutamine synthetase (GLNA 52 kDa), 50S ribosomal protein L2 (RL2 30 kDa), outer membrane protein A (OMPA 37 kDa) and transaldolase B (TALB 35 kDa).

Tab. 4.2. Carbonyl content in cell extracts of unirradiated and irradiated samples with a dose corresponding to less than 1h of natural sunlight irradiation (see Fig. 4.7A for corresponding blot). Computerized densitometry was done on films exposed for 30 sec to the membrane during chemiluminescent detection of carbonyl groups. Afterwards the density per ug of loaded protein and the increase in each sample compared to the first unirradiated control cell extract were calculated.

fluence (kJ m^{-2})	amount of protein (ug)	density (OD mm^{-2})	density/protein (OD mm^{-2} ug^{-1})	increase (fold)
0	7.5	29	3.86	**1**
0 (end)	7.5	31.13	4.15	**1.08**
215	7.5	52.08	6.94	**1.8**
430	7.5	53.9	7.19	**1.86**
645	7.5	103.92	13.86	**3.59**

Tab. 4.3. Carbonyl content in cell extracts of unirradiated and irradiated samples corresponding to about 5h of natural sunlight irradiation (see Fig. 4.7B for corresponding blot). Computerized densitometry was done on films exposed for 30 sec to the membrane during chemiluminescent detection of carbonyl groups. Afterwards the density per ug of loaded protein and the increase in each sample compared to the first unirradiated control cell extract were calculated.

fluence (kJ m^{-2})	amount of protein (ug)	density (OD mm^{-2})	density/protein (OD mm^{-2} ug^{-1})	increase (fold)
0	14.79	53.25	3.6	**1**
0 (end)	15.22	57.11	3.75	**1.04**
950	15.02	132.52	8.82	**2.45**
1900	14.9	211.36	14.19	**3.4**
2500	7.35	193.25	26.29	**7.3**

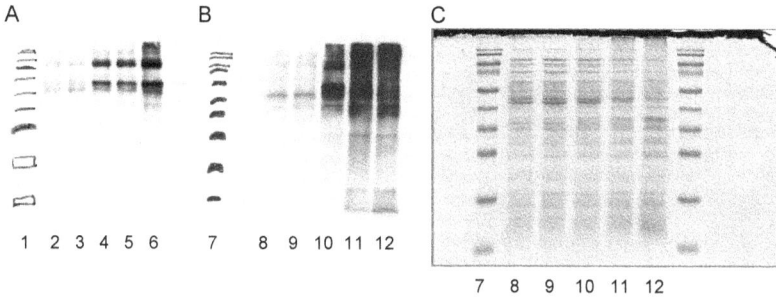

Fig. 4.7. Specific carbonylation of distinct protein bands is happening very early during irradiation at molecular weight ranges around 78 kDa, 55 kDa and 35 kDa. A: Oxyblot on whole cell extract with low fluences. Protein standard ladder (lane 1; 170, 130, 95, 72, 55, 43, 34, 26, 17, 10 kDa), dark controls (lanes 2&3), 215 kJ m^{-2} (lane 4), 430 kJ m^{-2} (lane 5), 645 kJ m^{-2} (lane 6) B: Oxyblot on whole cell extract with high fluences. protein std (lane 7), dark controls (lanes 8&9), 950 kJ m^{-2} (lane 10), 1900 kJ m^{-2} (lane 11), 2500 kJ m^{-2} (lane 12), C: 1D-SDS PAGE of samples shown in B. Experiments were done at least three times, representative results are shown.

Chapter 4

Discussion

Protein oxidation and carbonylation patterns. There are many different pathways that introduce carbonylation sites into proteins (Levine & Stadtman, 2001). The extent of protein carbonylation in biological samples is indicative for the level of oxidation processes that happen to cellular protein during a certain treatment or senescence. Oxidized proteins are non-functional and lose their structural or enzymatic function. It is believed that a high level of oxidized proteins causes substantial disruption of cellular functions. Carbonylated proteins can not be repaired, and have to be degraded. If not degraded immediately, they tend to form covalent crosslinks with other proteins and finally aggregate. Therefore, the increased smearing on immunoblots for fluences > 950 kJ m^{-2} could either originate from only a few proteins that were specifically carbonylated at lower fluences and then spread over a wider molecular weight range prior to blotting due to aggregation, or from a diversification of protein carbonylation targets with increasing irradiation dose.

A number of authors have reported increased carbonylation levels with aging, starvation and different stresses. Like in the case of UVA irradiation, these do not affect the proteome uniformly. Some of the proteins damaged during senescence and oxidative stress were found as major targets in other studies, too. For example, elongation factor G, glutamine syntethase, 60kDa chaperonine (Dukan & Nystrom, 1998; Dukan & Nystrom, 1999; Tamarit et al., 1998a), and outer membrane protein A and ATPase beta chain were reported to be affected, which corresponds to our observation on probable carbonylation candidates (Cabiscol et al., 2000). This suggests that the oxidative damage to the proteome is quite specific in all cases of increased oxidative stress and senescence, and again indicates that UVA light is effective via production of oxidative radicals and causes oxidative stress to the cell. Still, it is not clear how the level of carbonylation affects the viability of cells. Yeast cultures were reported to survive up to 10% of their proteome being irreversibly oxidized (Mirzaei & Regnier,

2006a). But dependent on how essential the function of a targeted protein is, e.g., if there are no isoenzymes that can replace it in the meantime for its function, carbonylation may be lethal. With essential functions of the energy metabolism targeted during SODIS (ATPase subunit alpha & beta) and the targets in the translation apparatus (elongation factor G, lysyl-tRNA synthetase, 50s ribosomal protein L2), a loss of viability due to carbonylation is probable. Protein modification is often accompanied by other cellular damages. For example in brain cell mitochondria, where respiration was inhibited, membrane potential and enzyme activities were reduced, ROS levels were elevated and the antioxidant defence system was decreased (Long *et al.*, 2009). It is likely that these effects influence each other, leaving us with the crucial question for what is the causative elicitor for the cellular damages, or in other words which one was the "chicken" and which ones are just „eggs". We believe that protein oxidation might be the elicitor for further cellular effects in the case of UVA irradiation, simply because it was observed very early during the inactivation process.

Aggregation (and in some cases fragmentation) of proteins during SODIS. It is widely accepted that enzymes with active site iron-sulfur clusters are highly sensitive to inactivation by oxygen radicals (Gardner & Fridovich, 1991). One of these enzymes is aconitase of *E. coli*, involved in iron pool regulation within the cell (Varghese *et al.*, 2003). Indeed, we do find this enzyme aggregated in our experiments. This again corroborates the presumption that UVA light acts via reactive oxygen. Aggregation is a direct consequence of protein oxidation. Our data clearly demonstrate that many different proteins within the cell are heavily affected by aggregation at an irradiation dose of about 1000 kJ m^{-2}, which corresponds to about two hours of natural sunlight and a 99% reduction when plating the cells on TSA agar. The capacity of the cellular quality control system by chaperones, therefore, seems to be exceeded at this point. Depletion of the cell from ATP earlier in the process (Berney *et al.*, 2006a; Bosshard *et al.*, 2009a) may be one cause for ATP-dependent chaperones not being able anymore to successfully fight protein misfolding and aggregation. For covalent protein crosslinking, protein oxidation damages (such as carbonylations) are an indispensable prerequisite. With such a wide variety of targeted proteins, impairing many different physiological functions within the cell, and such a high quantity of protein aggregates as we observed in our experiments, it is hard to imagine that cells could recover from the UVA induced damage. Nevertheless, some authors propose that initially, protein aggregation may be cytoprotective by sequestering non-functional proteins away from the cellular metabolism at specific sites and thereby also facilitating the recruitment of components of the cellular defence response (Maisonneuve et al., 2008c). Moreover, a small amount of aggregating proteins were found even in exponentially growing (so to say „young") cells, which means that protein aggregation per se is not lethal as long as it does not exceed the cells capacity to deal with the phenomenon. But the level of damaged and aggregated proteins significantly increased in aging bacterial populations during stationary phase and extensive protein aggregation might impair cell viability, e.g., by trapping other molecules (Wickner et al., 1999). With a wide variety of vital cellular functions hampered by aggregation, many secondary effects on cell physiology can be expected. Particularly important are probably the transcription and translation apparatus, transport systems, amino acid synthesis and degradation, respiration, ATP synthesis, glycolysis, the TCA cycle, chaperone functions and catalase. It is very likely that the lack of these functions directly impairs cellular viability during UVA irradiation.

Many enzymes may also be affected at fluences higher than 1000 kJ m^{-2}. For example, glyceraldehyde-3-P dehydrogenase was recently used as a model for environmentally induced protein damage (Voss et al., 2007). These authors found that UVA decreased the free thiol content, which indicates that the intracellular redox-balance was disturbed. The enzymes activity consequently decreased to 10% with about 1000 kJ m^{-2}, which is comparable to our results (Bosshard et al., 2009b). Aggregation and fragmentation effects in the same time were found only with fluences >2000 kJ m^{-2}. This is comparable to our results presented here, where aggregation of glyceraldehyde-3-P dehydrogenase was only seen at a fluence corresponding to about 2000 kJ m^{-2} (data not shown), but not at 1000 kJ m^{-2}. But since culturability of cells shows a massive drop at about 1000 kJ m^{-2}, applying higher fluences might be beating a dead horse. Other authors described that application of a lethal UVA dose shifted the location of HPI and HPII on gels, likely as cause of aggregation (Hoerter *et al.*, 2005a).

Defence against protein carbonylation and aggregation by chaperones and degradation enzymes. Proteins in a cell are affected by reversible and irreversible damage. Unwanted disulfide bond formation („disulfide stress") influences the folding of proteins within the cell, but it is reversible by gluthatione and chaperones. Irreversible is the oxidation of amino acid residues by metal ion-catalyzed oxidation reactions, leading (among other modifications) to protein carbonylation. Protein oxidation seems to enhance the ability of the cell to selectively degrade the affected protein. Therefore, some authors argued that carbonylation might be tagging a protein as a sign to the degradation machinery (Nyström, 2005). It was reported that *E. coli* has specific proteases that selectively degrade oxidized proteins in a ATP independent pathway (Davies & Lin, 1988). In this way the cells can circumwent the problem of ATP depletion in many situations of physiological stress. In contrast to this, chaperone systems usually depend on ATP for functioning, such as the DanK/DnaJ/GrpE system that helps most proteins during folding. For *E. coli* exposed to oxidative stress, which was associated with ATP depletion, Hsp33 was induced and replaced the DnaK system (Winter *et al.*, 2005). In this situation, DnaK was inactivated because its N-terminus becomes labile. Other authors claimed that Hsp33 is induced by disulfide formation under oxidative stress (Jakob *et al.*, 1999; Jakob *et al.*, 2000). Either way, Hsp33 was not induced to measurable amounts in our experiments because we worked with starved cells that are unable to induce much de-novo protein synthesis. Therefore, the DnaK system is still the more relevant in our case. Since a massive ATP depletion during UVA treatment was observed, it seems logical that DnaK and its substrate proteins (Winter *et al.*, 2005) are targeted by aggregation processes, and this is indeed what we were able to observe here. We found 11 of 39 DnaK substrate proteins aggregated, namely aldehyde-alcohol dehydrogenase, outer membrane protein A, aconitate hydratase 2, threonyl-tRNA synthetase, transketolase 1, pyruvate dehydrogenase, 2-oxoglutarate dehydrogenase E1 component, succinate dehydrogenase flavoprotein subunit, transcription termination factor rho, cysteine desulfurase, 3-oxoacyl-(acyl-carrier-protein) synthase 1. Also DnaK itself and other chaperones (ClpB, HtpG) were found to aggregate, and one must expect that this will even accelerate the process of protein aggregation during UVA treament.

Consequences of observed protein damages for the cell. Carbonylation and aggregation of proteins affect a wide spectrum of structural and enzymatic proteins within the cell. Vital cellular functions like the transcription and translation apparatus, transport systems, amino acid synthesis and degradation, respiration, ATP synthesis, glycolysis, TCA, chaperone functions and catalase are targeted by UVA irradiation. With the loss of catalase function, the cell loses its first line of defence against ROS, rendering the cell more susceptible against oxidative stress. A second line is lost by non-functional chaperones (be it due to lack of ATP to fuel chaperone function, be it due to aggregation of chaperones), which would be able to prevent protein aggregation. Moreover, damage to translational proteins reduces the cells ability to replace damaged proteins with new ones. The translation apparatus was a target of aggregation effects also in other studies (Mirzaei & Regnier, 2005; Mirzaei & Regnier, 2006a; Mirzaei & Regnier, 2006b; Mirzaei & Regnier, 2007). The proteins in the translation apparatus are very closely associated and therefore might be especially susceptible to aggregation. Moreover, an important protein for DNA protection and repair, Dps, was found to be damaged in our study. Dps (DNA protection under stationary phase) usually is induced with nutritional stress during stationary phase and under oxidative stress. It carries out two functions: it physically protects DNA from oxidative damage, additionally it maintains a low level of gene expression. Secondly, the structure of Dps is similar to ferritin and it was suggested that iron sequestration helps in the mechanism of protecting the DNA

(Grant *et al.*, 1998). Hence, its close association with iron might enhance photo-Fenton reactions during UVA irradiation and aggregation of the protein.

Accelerated senescence due to UVA stress. It was a long-standing belief that bacteria are not aging and that growing bacterial populations are not age structured. However, it has been recognized in the last years that even if bacterial populations are able to grow „eternally", individual bacterial cells indeed are aging, i.e. they slow down in their ability to divide and finally stop reproduction (Ackermann *et al.*, 2003; Lindner *et al.*, 2008; Nyström, 2007; Stewart *et al.*, 2005). Signs of aging have also been observed in stationary phase cells. The carbonylation pattern of non-growing cells shows that the cells get disfunctional in many respects (Desnues *et al.*, 2003; Nyström, 2003; Nyström, 2005; Nyström, 2006; Nyström, 2007). In aging cells, at least one out of every three proteins carries a carbonyl group, and this level of dysfunctional proteins is believed to be sufficient that carbonylation can be considered not only as a marker of aging, but also as a cause for the cellular changes during aging (Levine & Stadtman, 2001; Terman & Brunk, 2006). The pattern of carbonylation during aging reminds of the pattern we observed with UVA treatment. Accumulation of protein oxidation and protein aggregates has been described as a sign and a cause for cellular aging by many authors (Grune *et al.*, 2004; Maisonneuve *et al.*, 2008a; Stadtman, 2006; Terman & Brunk, 2006). We believe that UVA damages probably go back to a very similar cell damage mechanism also met during aging and oxidative stress in cells. UVA irradiation probably accelerates the processes observed during cellular aging.

Chapter 4

Acknowledgements

This project was financially supported by the Velux Foundation (project nr. 346). We thank Marc Suter, Holger Nestler and Victor Nesatyy for preliminary MS/MS experiments and valuable discussions about tandem mass spectrometry techniques and detection of post-translational modifications and the Funcional Genomics Center Zurich (fgcz) for measuring our samples and providing us with appropriate and up-to-date software for data evaluation. Ása Frederiksson and Thomas Nyström generously supported us with their experience and knowledge while establishing the Oxyblot method in our laboratory.

5 Conclusions, outlook and implications for the SODIS method

Cellular energy metabolism is the key target of SODIS

Cell inactivation during SODIS is a fact, but not much was known about the underlying inactivation mechanisms. In order to find the basic mechanisms of inactivation, the focus of this thesis was put on cellular targets that are affected early during irradiation and are essential in the metabolism. The question which essential pathways lead to die-off can now be answered with considerable certainty. The crucial damages during SODIS target the cellular energy generating systems (Fig. 5.1). First changes were observed as a massive reduction of enzymatic activities of the respiratory chain already at fluences around 50 kJ m^{-2}, corresponding to less than 10 min. of sunlight. The fact that all respiratory dehydrogenases tested in this thesis (NADH, succinate, lactate DH) are inactivated at a similar fluence suggests that these enzymes are not inactivated independently, but that the electron transport chain itself is interrupted by UVA irradiation. The cellular ATP pool is exhausted at around 300 kJ m^{-2}, followed by an inactivation of the ATPase at around 900 kJ m^{-2} (Bosshard et al., 2009b). These processes are very likely accompanied by a slow but continuous loss of membrane potential, which is completed at approximately 1900 kJ m^{-2}. Efflux pumps directly depend on the membrane potential and probably as a consequence of the loss of membrane potential, efflux pump activity stops at around 900 kJ m^{-2} (Berney et al., 2006a). The membrane potential might be kept up for some time by the reverse use of the ATPase and this could explain that the efflux pump activity is kept up approximately as long as the ATPase is still active, as discussed in (Bosshard et al., 2009b). As an example for a system not dependent on membrane potential, glucose uptake is fully active up to a fluence of around 800 kJ m^{-2}. Then it slowly drops down in activity until it shuts down at a fluence of 1500 kJ m^{-2} (Berney et al., 2006a), demonstrating that this system sustains relatively long. The ability of E. coli cells to generate ATP when nutrients are available (ATP generation potential ATP-GP) becomes gradually restricted with irradiation doses higher than 180 kJ m^{-2}. Cells are unable to recover once irradiated with fluences higher than 300 kJ m^{-2}, meaning that inactivation continues even in the absence of light (Bosshard et al., 2009b). To summarize the effect of UVA irradiation on different cellular functions, F90-values, i.e. the fluence at which 90% of the function is lost, were calculated for all measured parameters in E. coli (Tab. 5.1).

Tab. 5.1. Summary of measured cellular functions in E. coli and fluences at which the enzyme had lost 90% of its activity or cell damage had reached 90% of the endpoint (F90 values)

cellular function	F90 (kJ/m^2)	reference
respiratory chain activity	50	(Bosshard et al., 2009b)
ATP generation potential (ATP-GP)	180	(Bosshard et al., 2009b)
ATP per cell	300	(Berney et al., 2006a)
culturability followed by dark storage for 48h ("point of no return")	300	(Bosshard et al., 2009b)
hydroperoxidases activity	400	(Bosshard et al., 2009b)
specific protein carbonylation	600	(Bosshard et al., 2009c)
ATPase activity	900	(Bosshard et al., 2009b)
efflux pumps	900	(Berney et al., 2006a)
culturability during continuous irradiation	900	(Berney et al., 2006a; Bosshard et al., 2009b)
cytoplasmatic enzymes	>1000	(Bosshard et al., 2009b)
glucose uptake (2-NBDG)	1500	(Berney et al., 2006a)
membrane polarisation	1900	(Berney et al., 2006a)

membrane integrity	2400	(Berney *et al.*, 2006a)
protein aggregation	2400	(Bosshard *et al.*, 2009c)

In further research, uptake kinetics of defined single substrates could reveal in more detail, which substrate uptake systems become damaged by UVA irradiation. Further, a closer look at cell viability and the ATP generation potential (ATP-GP) would be worthwhile. Hitherto the definition of cell viability is very ambiguous. A cell that is able to grow on plates is traditionally judged as viable in microbiology (Kell *et al.*, 1998). However, many microbiologists share the opinion that a large part of the viable cells are not able to grow on agar plates since the "unnatural" nutritional milieu does not favor growth (Colwell *et al.*, 1985; Roszak & Colwell, 1985). Closer to natural conditions is the approach where diluted nutrients are added to cells and cell growth is then observed directly by microscopy, the so called DVC ("direct viable count") approach (Hoefel *et al.*, 2003; Kogure *et al.*, 1987). Irradiated cells could be tested in this way for cell growth and viability (Villarino *et al.*, 2000; Villarino *et al.*, 2003). The observed inhibitory effects on ATP-production (Bosshard *et al.*, 2009b), which is a bulk-parameter, could be interpreted on the single cell level.

Since the respiratory chain seems so crucial in the inactivation process, identifying the main target of irradiation would be important. There are several places where the electron transport chain can be interrupted by UVA-specific damage. Therefore, different components of the respiratory chain would have to be tested by adding specific substrates and inhibitors in continuative experiments. The electron transport chain in mitochondria has been widely investigated in connection with oxidative stress. It was found that in the mitochondrial electron transport chain, the flavin mononucleotide group of complex I and not the ubiquinone of complex III is the primary physiological relevant site of reactive oxygen species (ROS) generation (Yuanbin *et al.*, 2002). The electron transport chain of *E. coli* has not been examined in that respect so far, but since electron transport chains are evolutionarily quite conserved, also in the respiratory chain of *E. coli* flavin enzymes are suspected as primary target of ROS damage during UVA irradiation.

Cell physiology is affected by protein carbonylation and aggregation damages

The causative agents for the "toxicity" of light seem to be ROS, generated, e.g., by photo-Fenton reactions in membranes. Especially the hydroxyl radical is highly reactive and damages proteins in many different oxidative ways (Imlay, 2009). In general, the respiratory chain in biological membranes is producing the lion's share, around 90%, of ROS within cells (Cabiscol *et al.*, 2000). Biological membranes contain many photo-sensitizers that are able to produce ROS via photo-Fenton reactions. The high content of photo-sensitizers is evident when biological membranes are purified, e.g., by differential centrifugation, simply because the membrane fraction appears slightly red-colored.

Oxidative damage at the level of proteins was observed in this thesis (i) as carbonylation products, which were evident early, starting from approximately 200 kJ m^{-2}, and reaching a maximum at a fluence of about 600 kJ m^{-2} and (ii) as increased protein aggregation as a consequence of oxidative damage to cellular proteins. Protein aggregation was detected later in the process of inactivation, starting at approximately 500 kJ m^{-2} and increasing thereafter (Fig. 5.1). At a fluence of 1000 kJ m^{-2}, more than one third of the identified proteins were affected by aggregation. Therefore, protein damages seem to be important for the breakdown of many physiological functions during SODIS, including the transcription and translation apparatus, transport systems, amino acid synthesis and degradation, respiration, ATP synthesis, glycolysis, TCA cycle, chaperone functions and catalases (Bosshard *et al.*, 2009c). The effects of SODIS on cell physiology and proteome are summarized in a flow-diagram (Fig. 5.2). There is a lot of literature on oxidative stress, senescence and protein damages. However, there is no "final answer" whether the protein damages found in our study are the most important cause for the loss of cell viability or only a consequence thereof.

Therefore, it would be interesting to see whether or not the protein damages found in this thesis are hampering cell viability by employing assays that quantify the activity of the individual cellular systems, e.g., translation by radio-labeled leucine incorporation (Kirchman *et al.*, 1985), or transcription by radio-labeled thymidine incorporation (Daneri *et al.*, 1994). Further, protein aggregates could be visualized *in vivo* by electron microscopy methods.

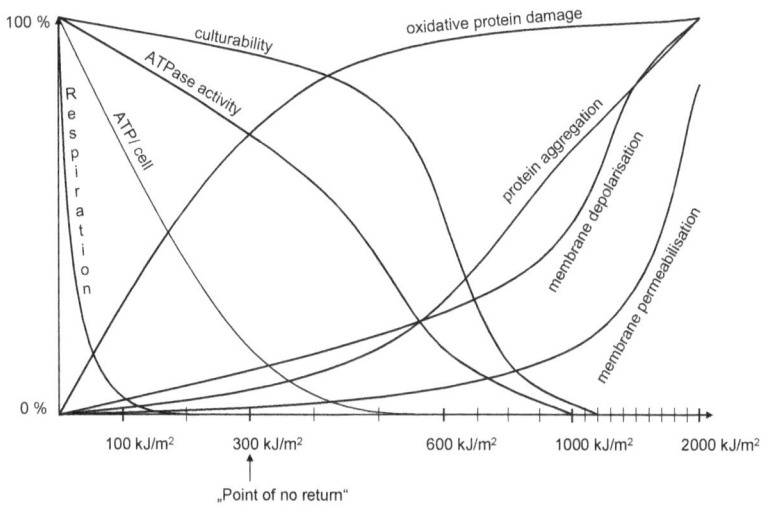

Fig. 5.1. Summary of observed sequential breakdown of cellular functions in *E. coli* during solar disinfection with approximate fluences. One tick on the x-axis corresponds to 100 kJ m^{-2}.

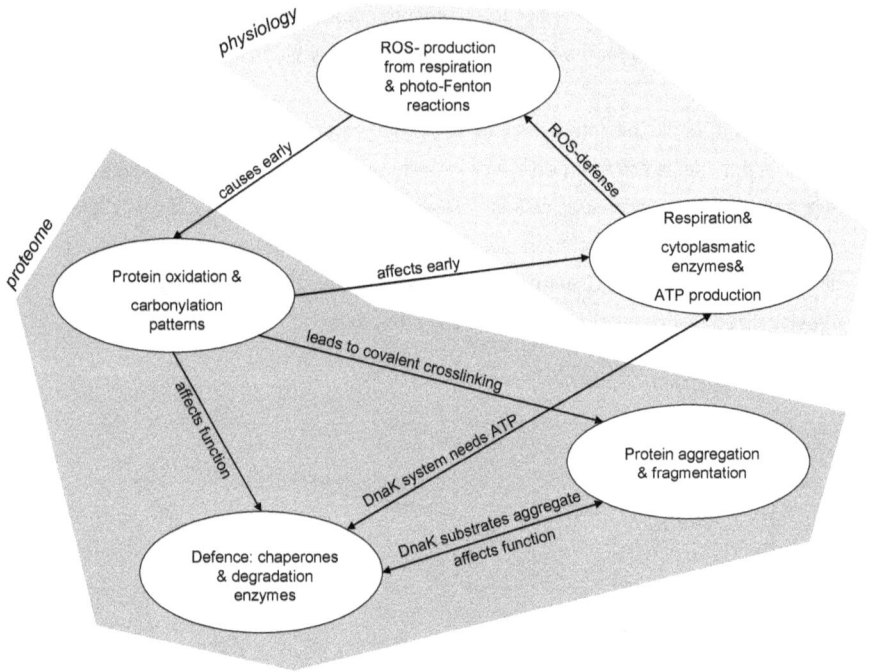

Fig. 5.2. Overview of effects of sunlight irradiation on cellular physiology in bacteria, damaged cellular systems and their relationship.

As a further research area, it would be interesting to have a closer look specifically at membrane proteins and their damage levels. This was tried for fractionated cells using the oxyblot technique during this thesis. It appears that the cellular fractions were not pure enough, so that the same bands of proteins with carbonylation damage were found in the cytoplasm as well as in the inner membrane fraction. However, the approach to identify aggregated proteins via one-dimensional SDS PAGE (Bosshard et al., 2009c) could be used specifically for membrane proteins, too. Unfortunately, other proteomics techniques such as 2D-gel electrophoresis are difficult to perform on membrane proteins because of their lipophilic character and subsequently low solubility under conditions suitable for 2D-gels (Rabilloud, 2009; Santoni *et al.*, 2000). Besides this, we encountered serious difficulties when trying to resolve aggregated proteins with 2D-gel electrophoresis. This could probably be circumvented by using cells that are irradiated only to a level where oxidative carbonylation damage becomes evident, and before extensive aggregation has taken place. 2D-gel electrophoresis could then be combined either with staining of proteins with carbonylation damages by Avidin-FITC, or carbonylated proteins could directly be enriched from whole cell extracts or cell fractions by affinity enrichment (Mirzaei & Regnier, 2005; Mirzaei & Regnier, 2007). It would also be interesting not only to identify the individual proteins that are damaged, but to find out whether certain amino acids are targeted specifically by UVA irradiation. A recently published paper states that the neighboring amino acids within a protein are determinants for amino acid damage by oxidative stress (Maisonneuve *et al.*, 2009).

Protein damages are a likely cause for cellular die-off, even more than DNA damages

Effects observed early during solar irradiation at the physiological level, i.e. breakdown of respiratory chain, ATPase activity and membrane potential, and later protein aggregation are most likely the result of one and the same cause: the oxidative damage to proteins in membranes. DNA damage was also discussed before as a cause for cellular die-off during UVA irradiation (Jiang *et al.*, 2009). However, it is likely that damage to proteins might hurt the cells more severely, because the damages are happening early and in systems that are indispensible for cell viability. DNA damage would become important for the cells especially when trying to synthesize new proteins or to proliferate. As long as the cells stay in a "dormant" state (in which they were in our experiments, because they stayed in stationary phase without nutrients), DNA damages would not be as crucial as protein damages. Moreover, as a consequence of protein damage, the cell would not be able to repair DNA damages anymore because it lacks appropriate energy supply from ATP. The avoidance of this chain of effects was also recently suspected to be important in x-ray resistance of *Deinococcus radiodurans*, which is extremely radiant resistant probably due to an efficient system protecting its proteins from oxidative damage (Daly *et al.*, 2007). It was suggested that *D. radiodurans* protects its proteins from oxidative damage by an unusually high Mn/Fe ratio in the cytoplasm (Daly *et al.*, 2004). Mn seems to prevent oxidative damage by scavenging ROS. These observations combined with our observations on extensive protein damages in UVA-treated cells let us conclude that protein damage might be more important for cell viability during UVA irradiation than DNA damage.

Further research on this topic should test whether an increased cytosolic Mn or decreased Fe content could influence UVA susceptibility also in cells of *E. coli*. Comparison of the resistance mechanism of *D. radiodurans* with other bacteria that are resistant to UVA light, e.g., *S. typhimurium* could give a hint if there is a general mechanism for radiation resistance via scavenging of ROS. To compare differences at the molecular level, proteomic analysis could be combined with transcriptomics and whole-genome comparison approaches. Furthermore, mutants from a commercially available *E. coli* mutant library (Keio collection) could be studied to confirm the crucial role of proteins. UVA susceptibility of mutants that lack a proper DNA repair machinery (SOS response, excision repair), systems to protect proteins or DNA (chaperones, heat shock proteins, dps), components of the stringent response and enzymes that specifically protect cells from oxidative stress (hydroperoxidases, superoxide dismutases) should be tested.

Accelerated aging in bacterial cells due to SODIS

In the recent years, the mechanisms of cell aging have gained increasing attention in the scientific community. Signs of cellular aging have not only been observed in multicellular organisms, but also in bacterial cells that were supposed to be "immortal". One single bacterium that was observed for its capability for division over time, slowed down in its division rate until it stopped dividing (Ackermann *et al.*, 2003; Nyström, 2007). Thus, bacteria show a reduced reproduction ability when they get older. Moreover, there seems to be a differentiation into a "mother" and a "daughter" cell by an asymmetrical inheritance of the cell material during division. The daughter cell is built from new material and, therefore, starts with an intact reproduction potential. Not much is known about the mechanisms of cellular aging, but proteins seem to play a crucial role. Aging cells were found to contain considerably more protein aggregates and oxidative protein damages than juvenile ones (Maisonneuve *et al.*, 2008a; Maisonneuve *et al.*, 2008b; Maisonneuve *et al.*, 2008c). From the mechanistic side, there are three hypothetical underlying mechanisms leading to cell aging.

- "garbage" theory: stochastic deterioration in the cell leads to an accumulation of cellular "garbage" (i.e. protein aggregates), more and more disrupting proper cell functioning and finally leading to cell death (Terman & Brunk, 2006).

- "programmed cell death" or "bacterial apoptosis" theory: genetically programmed, well organized process comparable to apoptosis in eukaryotic cells. This voluntary "suicide" is argued to have social function in the bacterial community since nutrients from the dead cell get available for the cells neighbors(Aizenman *et al.*, 1996). However, this theory is highly controversial (Nyström, 2007).

- "organized garbage" theory: some authors argue that the cell inevitably accumulates damaged proteins when it gets older. To circumvent the situation that the whole cell is filled up with aggregated proteins at some point, the cell "directs" the garbage to one "trash organelle" in the cell, in order to keep up cell functions as long as possible (Maisonneuve *et al.*, 2008c; Nyström, 2007). Asymmetric inheritance of cell material during cell division was observed before and is a very likely reason for the aging of the "mother cell" (Ackermann *et al.*, 2003; Nyström, 2007).

The "garbage" theory is the most likely mechanism during SODIS. The cells are left very soon without intracellular energy and, therefore, are unable to direct "garbage", i.e. protein aggregates, to a specific location. SODIS seems to accelerate processes that usually are happening during bacterial cell-aging (Bosshard *et al.*, 2009c).

Post-irradiation inactivation processes

For *E. coli* and *Salmonella typhimurium*, dark inactivation after UVA irradiation when storing the cells in nutrient poor water was observed. These post-irradiation processes start to take place already after exposure to low fluences thus enhancing SODIS and shortening the times of exposure needed. In *S. typhimurium*, the measured viability indicators were inactivated with a much lower fluence than expected when an overnight break was introduced during the exposure process (Bosshard *et al.*, 2009a). These results indicate that SODIS becomes more effective when the fluence is fractionated. In *E. coli*, a fluence of approximately 300 kJ m^{-2} (one third of the fluence needed for loss of culturability, Tab. 5.1) is needed before dark inactivation processes commence and the "point of no return" is reached. The reason for further die-off in the dark might be ATP depletion, as discussed in (Bosshard *et al.*, 2009a), and ongoing damages to essential structures and metabolic pathways of the cell. The cell might even get unable to repair proteins by chaperones because it lacks appropriate energy provision for chaperone action, like DnaK (Winter *et al.*, 2005), as shown in Fig. 5.2. The aftereffects of irradiation may be due to radical chain reactions of lipids and proteins. Since bacteria are not susceptible for lipid peroxidation because of their membrane composition (Imlay, 2009), proteins in this case are better candidates for chain reactions. Radical chain reactions have been well documented for proteins (Berlett & Stadtman, 1997; Dean *et al.*, 1997). Even if nutrients were supplied, cells were unable to efficiently generate ATP (Bosshard *et al.*, 2009a; Bosshard *et al.*, 2009b). Therefore, the protein damages to the cells by UVA light seem to be irreparable. The cells can not escape and thus are prone to die already after relatively short irradiation periods. This means that SODIS-treated bacteria will not have a chance for recovery, even if they are in a nutrient rich environment. This is relevant for practice, where sometimes water with a high organic load is SODIS-treated.

Relevance for practice

Since SODIS is used in practice to improve the microbiological drinking water quality, transferability of academic research to practice has gained more and more attention in the recent years. Some relevant points for practice are discussed here:

- **Applicability of results achieved with artificial light compared to exposure to sunlight**

Results obtained in laboratory experiments were highly similar to results from cells irradiated with sunlight, as long as the light intensity in the laboratory was varied only within the normal range of sunlight conditions (Berney *et al.*, 2006a; Berney *et al.*, 2006d; Bosshard *et al.*, 2009a). So called "dose reciprocity law failures" manifested themselves primarily for irradiation intensities exceeding sunlight intensity, where cells were damaged with a much lower dose than expected. Quality and quantity of cell damages in sunlight monitored by viability indicators, ATP measurements and culturability corresponded very well with those obtained with artificial light under laboratory conditions. This was true even for two different types of medium pressure mercury lamps, although the wavelength spectrum of these lamps is different from sunlight. This suggests that the "active wavelengths" in sunlight that cause cell damage are also included in the spectrum of the mercury lamps used (Fig. 5.3).

- **Transferability of our results to SODIS "real world" scenarios**

There are different issues that have to be accounted for to transfer the knowledge obtained in this thesis to the situation in the field. Many parameters that influence cell damage are controlled easily in irradiation experiments performed at our institute, such as water temperature, water turbidity, chemical water quality in the source water, organic load in the source water, and concentration and growth state of bacteria in the source water. These factors are often unpredictable in the field. Solar light and artificial UVA disinfection was studied in this thesis at 37°C and in waters supporting proper cell physiology.

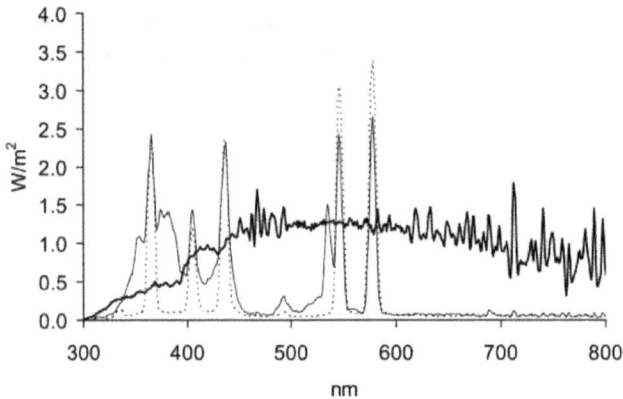

Fig. 5.3. Wavelenght spectra of medium pressure mercury lamps Hanau TQ718 (dashed line) and TQ718 Z4 (thin solid line) both corrected for filter solution $NaNO_3$ and the sunlight spectrum on a mid-summer day measured at 1 pm (fat solid line). (Berney et al., 2006b)

Increased water temperatures lead to a synergistic effect for microbial die-off (Joyce et al., 1996; Wegelin et al., 1994). Water turbidity definitely decreases effective application of SODIS in practice. Also poor bottle maintenance resulting in dirty, scratched bottles can decrease the light dose and promote recontamination of the treated water. The organic load of natural source waters varies in a wide range. Since nutrient addition is not helping bacteria to recover after irradiation (Bosshard et al., 2009b), it is assumed that even a good nutrient supply would not protect cells from the deleterious effects of sunlight. Anyway, nutrients in natural waters usually are very limited and, therefore, most bacteria are in stationary phase or growing very slowly only (Moriarty & Bell, 1993). A worst case scenario was used in our experiments by working with stationary phase cells, the most UVA-resistant growth state (Berney *et al.*, 2006b). Therefore, disinfection efficacies under laboratory and field conditions match well.

Nevertheless, epidemiological field studies on the health impact of SODIS lead to controversial results. The disinfection efficacy of SODIS was well confirmed in the laboratory and in the field, but does not translate easily into a significant health impact. While most authors are very convinced that application of SODIS improves the health situation in practice (Conroy *et al.*, 1996; Conroy *et al.*, 2001; Rose *et al.*, 2006), others have questioned the positive impact (Hunter, 2009; Mäusezahl *et al.*, 2009; Sobsey *et al.*, 2008). The most probable reason for these divergent results for the health impact of SODIS is that many socio-cultural issues influence the compliance of SODIS users. A health impact is seen only when household water treatment, such as SODIS, is practiced consistently and only safe water is consumed. This fact stresses the importance of good diffusion and training strategies to distribute and anchor the technique in local communities (Altherr *et al.*, 2008; Heri & Mosler, 2008). Moreover, pathogens causing diarrhoea are transmitted through multiple pathways, such as contaminated water and food, improper hygiene practices and inadequate sanitary conditions. Therefore, diarrhoeal illnesses can not be eradicated with interventions targeting safe water only.

- **Could SODIS efficiency be enhanced for practice?**

The idea to enhance the disinfection efficiency of SODIS for the field application is appealing, but also challenging, since one of the main advantages of SODIS is its simplicity. Therefore, approaches to enhance SODIS should aim at keeping the method as simple as possible. One very simple way is a fractionation of the fluence that is applied. An interruption of sunlight exposure over night resulted in a reduction of the overall fluence for the same disinfection effect (Bosshard *et al.*, 2009a). It is very well possible that fluence fractionation takes place in practice unintentionally, since people are advised to expose the water on two consecutive days in case of overcast conditions. Furthermore, high concentrations of humic acids in the source water were described to enhance solar inactivation of viruses. The mechanism behind this effect is most probably a close association of humic acids to the virus and the production of singlet oxygen (Kohn *et al.*, 2007; Kohn & Nelson, 2007). However, an addition of humic acids in the field is not feasible. Other additives are more likely to come into consideration. TiO_2 was applied as a catalyst (Mendez-Hermida *et al.*, 2007; Rincon & Pulgarin, 2004) and citric acid (from lime juice), hydrogen peroxide, copper and ascorbate were used as additives (Fisher *et al.*, 2008a; Fisher *et al.*, 2008b) to enhance the SODIS effect. The question if these methods could be distributed and used efficiently in practice still needs to be evaluated.

References

Ackermann, M., Stearns, S. C. & Jenal, U. (2003). Senescence in a bacterium with asymmetric division. *Science* **300**, 1920.

Acra, A. (1980). Disinfection of oral rehydration solutions by sunlight. *Lancet* **316**, 1257-1258.

Acra, A. (1989). Sunlight as disinfectant. *Lancet* **333**, 280.

Ahmad, S. I. (1981). Synergistic action of near ultraviolet radiation and hydrogen peroxide on the killing of coliphage T7: possible role of superoxide radical. *Photochem Photobiol* **2**, 173-180.

Aizenman, E., Engelberg-Kulka, H. & Glaser, G. (1996). An *Escherichia coli* chromosomal "addiction module" regulated by guanosine [corrected] 3',5'-bispyrophosphate: a model for programmed bacterial cell death. *Proc Natl Acad Sci USA* **93**, 6059-6063.

Allison, W. S. & Kaplan, N. O. (1964). The comparative enzymology of triosephosphate dehydrogenase. *J Biol Chem* **239**, 2140-2152.

Altherr, A. M., Mosler, H. J., Tobias, R. & Butera, F. (2008). Attitudinal and relational factors predicting the use of solar water disinfection: A field study in Nicaragua. *Health Educ Behav* **35**, 207-220.

Anderson, J. W., Foyer, C. H. & Walker, D. A. (1983). Light-dependent reduction of dehydroascorbate and uptake of exogenous ascorbate by spinach chloroplasts. *Planta* **158**, 442-450.

Arami, S. I., Hada, M. & Tada, M. (1997a). Near-UV-induced absorbance change and photochemical decomposition of ergosterol in the plasma membrane of the yeast *Saccharomyces cerevisiae*. *Microbiology* **143**, 1665-1671.

Arami, S. I., Hada, M. & Tada, M. (1997b). Reduction of ATPase activity accompanied by photodecomposition of ergosterol by near-UV irradiation in plasma membranes prepared from *Saccharomyces cerevisiae*. *Microbiology* **143**, 2465-2471.

Ascenzi, J. M. & Jagger, J. (1979). Ultraviolet action spectrum (238-405 nm) for inhibition of glycine uptake in *E. coli*. *Photochem Photobiol* **30**, 661-666.

Ballesteros, M., Fredriksson, A., Henriksson, J. & Nyström, T. (2001). Bacterial senescence: protein oxidation in non-proliferating cells is dictated by the accuracy of the ribosomes. *EMBO J* **20**, 5280-5289.

References

Berlett, B. S. & Stadtman, E. R. (1997). Protein oxidation in aging, disease, and oxidative stress. *J Biol Chem* **272**, 20313-20316.

Berney, M., Weilenmann, H.-U. & Egli, T. (2006a). Flow-cytometric study of vital cellular functions in *Escherichia coli* during solar disinfection (SODIS). *Microbiology* **152**, 1719-1729.

Berney, M., Weilenmann, H.-U., Ihssen, J., Bassin, C. & Egli, T. (2006b). Specific growth rate determines the sensitivity of *Escherichia coli* to thermal, UVA, and solar disinfection. *Appl Environ Microbiol* **72**, 2586-2593.

Berney, M., Weilenmann, H. U. & Egli, T. (2006c). Gene expression of *Escherichia coli* in continuous culture during adaptation to artificial sunlight. *Environ Microbiol* **8**, 1635-1647.

Berney, M., Weilenmann, H. U., Simonetti, A. & Egli, T. (2006d). Efficacy of solar disinfection of *Escherichia coli, Shigella flexneri, Salmonella* Typhimurium and *Vibrio cholerae. Appl Microbiol* **101**, 828-836.

Berney, M., Hammes, F., Bosshard, F., Weilenmann, H. U. & Egli, T. (2007a). Assessment and interpretation of bacterial viability by using the LIVE/DEAD BacLight Kit in combination with flow cytometry. *Appl Environ Microbiol* **73**, 3283-3290.

Berney, M., Weilenmann, H.-U. & Egli, T. (2007b). Adaptation to UVA radiation of *E. coli* growing in continuous culture. *J Photochem Photobiol B* **86**, 149-159.

Berney, M., Vital, M., Hülshoff, I., Weilenmann, H.-U., Egli, T. & Hammes, F. (2008). Rapid, cultivation-independent assessment of microbial viability in drinking water. *Water Res* **42**, 4010-4018.

Beyer, W. F. & Fridovich, I. (2002). Effect of hydrogen peroxide on the iron-containing superoxide dismutase of Escherichia coli. *Biochemistry* **26**, 1251-1257.

Bochner, B. R. (2009). Global phenotypic characterization of bacteria. *FEMS Microbiol Rev* **33**, 191-205.

Bosshard, F., Berney, M., Scheifele, M., Weilenmann, H.-U. & Egli, T. (2009a). Solar disinfection (SODIS) and subsequent dark storage of *Salmonella typhimurium* and *Shigella flexneri* monitored by flow cytometry. *Microbiology* **155**, 1310-1317.

Bosshard, F., Bucheli, M., Meur, Y. & Egli, T. (2009b). The respiratory chain is the cells Achilles' heel during UVA inactivation in *Escherichia coli* (submitted).

Bosshard, F., Riedel, K., Schneider, T., Geiser, C., Bucheli, M. & Egli, T. (2009c). Protein oxidation and aggregation in UVA irradiated *Escherichia coli* cells as signs of accelerated cellular senescence (submitted).

Bourdon, E. & Blache, D. (2001). The importance of proteins in defence against oxidation. *Antioxid Redox Signal* **3**, 293-311.

Bradford, M. M. (1976). A rapid and sensitive method for the quantitation of microgram quantities of protein utilizing the principle of protein-dye binding. *Anal Biochem* **72**, 248-254.

Cabiscol, E., Tamarit, J. & Ros, J. (2000). Oxidative stress in bacteria and protein damage by reactive oxygen species. *Int Microbiol* **3**, 3-8.

Chamberlain, J. & Moss, S. H. (1987). Lipid peroxidation and other membrane damage produced in *Escherichia coli* K1060 by near-UV radiation and deuterium oxide. *Photochem Photobiol* **45**, 625-630.

Chiti, F. (2006). Relative improtance of hydrophobicity, net charge and secondary structure propensities in protein aggregation. In *Protein misfolding, aggregation and conformational diseases; Part A: Protein Aggregation and conformational diseases*. Edited by V. N. Uversky & A. L. Fink: Springer.

Choksi, K. B., Nuss, J. E., DeFord, J. H. & Papaconstantinou, J. (2008). Age-related alterations in oxidatively damaged proteins of mouse skeletal muscle mitochondrial electron transport chain complexes. *Free Radical Bio Med* **45**, 826-838.

Claiborne, A. & Fridovich, I. (1979). Purification of the o-dianisidine peroxidase from *Escherichia coli* B. Physicochemical characterization and analysis of its dual catalatic and peroxidatic activities. *J Biol Chem* **254**, 4245-4252.

Claiborne, A., Malinowski, D. P. & Fridovich, I. (1979). Purification and characterization of hydroperoxidase II of *Escherichia coli* B. *J Biol Chem* **254**, 11664-11668.

Colwell, R. R., Brayton, P. R., Grimes, D. J., Roszak, D. B., Huq, S. A. & Palmer, L. M. (1985). Viable but non-culturable *Vibrio Cholerae* and related pathogens in the environment - implications for release of genetically engineered microorganisms. *Bio/Technology* **3**, 817-820.

Conroy, R. M., Elmore-Meegan, M., Joyce, T., McGuigan, K. G. & Barnes, J. (1996). Solar disinfection of drinking water and diarrhoea in Maasai children: a controlled field trial. *Lancet* **348**, 1695-1697.

References

Conroy, R. M., Meegan, M. E., Joyce, T., McGuigan, K. & Barnes, J. (1999). Solar disinfection of water reduces diarrhoeal disease: an update. *Arch Dis Child* **81**, 337-338.

Conroy, R. M., Meegan, M. E., Joyce, T., McGuigan, K. & Barnes, J. (2001). Solar disinfection of drinking water protects against cholera in children under 6 years of age. *Arch Dis Child* **85**, 293-295.

D'Alessandro, M., Turina, P. & Melandri, B. A. (2008). Intrinsic uncoupling in the ATP synthase of *Escherichia coli*. *BBA - Bioenergetics* **1777**, 1518-1527.

Daly, M. J., Gaidamakova, E. K., Matrosova, V. Y. & other authors (2004). Accumulation of Mn(II) in *Deinococcus radiodurans* facilitates gamma-radiation resistance. *Science* **306**, 1025-1028.

Daly, M. J., Gaidamakova, E. K., Matrosova, V. Y. & other authors (2007). Protein oxidation implicated as the primary determinant of bacterial radioresistance. *PLoS Biol* **5**, 769-779.

Daneri, G., Riemann, B. & Williams, P. J. I. (1994). In situ bacterial production and growth yield measured by thymidine, leucine and fractionated dark oxygen uptake. *J Plankton Res* **16**, 105-113.

Davies, K. J. (1987). Protein damage and degradation by oxygen radicals. I. general aspects. *J Biol Chem* **262**, 9895-9901.

Davies, K. J. & Delsignore, M. E. (1987). Protein damage and degradation by oxygen radicals. III. Modification of secondary and tertiary structure. *J Biol Chem* **262**, 9908-9913.

Davies, K. J. & Lin, S. W. (1988). Oxidatively denatured proteins are degraded by an ATP-independent proteolytic pathway in *Escherichia coli*. *Free Radical Bio Med* **5**, 225-236.

Dean, R. T., Fu, S., Stocker, R. & Davies, M. J. (1997). Biochemistry and pathology of radical-mediated protein oxidation. *Biochem J* **324**, 1-18.

Desnues, B., Cuny, C., Dukan, S., Grégori, G., Aguilaniu, H. & Nyström, T. (2003). Differential oxidative damage and expression of stress defence regulons in culturable and non-culturable *Escherichia coli* cells. *EMBO Rep* **4**, 400-404.

Dukan, S. & Nystrom, T. (1998). Bacterial senescence: stasis results in increased and differential oxidation of cytoplasmic proteins leading to developmental induction of the heat shock regulon. *Genes Dev* **12**, 3431-3441.

Dukan, S. & Nystrom, T. (1999). Oxidative stress defence and deterioration of growth-arrested *Escherichia coli* cells. *J Biol Chem* **274**, 26027-26032.

Dzidic, S., Salaj-Smic, E. & Trgovcevic, Z. (1986). The relationship between survival and mutagenesis in *Escherichia coli* after fractionated ultraviolet irradiation. *Mutat Res Lett* **173**, 89-91.

Eisenstark, A. (1970). Sensitivity of *Salmonella typhimurium* recombinationless (rec) mutants to visible and near-visible light. *Mutat Res* **10**, 1-6.

Evison, L. M. (1988). Comparative studies on the survival of indicator organisms and pathogens in fresh and sea water. *Water Sci Technol* **20**, 309-315.

Favre, A. & Hajnsdorf, E. (1983). Photoregulation of *E. coli* growth and the near UV photochemistry of tRNA. In *Molecular Models of Photoresponsiveness*. Edited by B. F. Erlanger & G. Montagnoly. New York: Plenum.

Favre, A., Hajnsdorf, E., Thiam, K. & Caldeira de Araujo, A. (1985). Mutagenesis and growth delay induced in *E. coli* by near-UV radiations. *Biochimie* **67**, 335-342.

Fisher, M. B., Keenan, C. R., Nelson, K. L. & Voelker, B. M. (2008a). Speeding up solar disinfection (SODIS): effects of hydrogen peroxide, temperature, pH, and copper plus ascorbate on the photoinactivation of *E. coli*. *J Water Health* **6**, 35-51.

Fisher, M. B., Nelson, K. L. & Iriarte, M. (2008b). Sunlight inactivation rates of laboratory-cultured and wastewater-derived *E. coli* and *Enterococci* in the presence and absence of additives and iron chelators: applied lessons for SODIS (unpublished).

Franchini, A. G. (2006). Physiology and fitness of *Escherichia coli* during growth in carbon-excess and carbon-limited environments. Zurich: ETH PhD-thesis Nr. 16585.

Fredrickson, J. K., Li, S. M. W., Gaidamakova, E. K. & other authors (2008). Protein oxidation: key to bacterial desiccation resistance? *Isme J* **2**, 393-403.

Friguet, B. (2006). Oxidized protein degradation and repair in aging and oxidative stress. *FEBS Letters* **30557**.

Fujioka, R. S., Hashimoto, H. H., Siwak, E. B. & Young, R. H. (1981). Effect of sunlight on survival of indicator bacteria in seawater. *Appl Environ Microbiol* **41**, 690-696.

References

Gallagher, S. R. (2006). One-dimensional SDS gel electrophoresis of proteins. *Curr Prot Immunol* **unit 8.4.**

Gameson, A. & Saxon, D. (1967). Field studies on effect of daylight on mortality of coliform bacteria. *Water Res* **1**, 279-295.

Gardner, P. R. & Fridovich, I. (1991). Superoxide sensitivity of the Escherichia coli aconitase. *J Biol Chem* **266**, 19328-19333.

Geeraerd, A. H., Valdramidis, V. P. & Van Impe, J. F. (2005). GInaFiT, a freeware tool to assess non-log-linear microbial survivor curves. *Int J of Food Microbiol* **102**, 95-105.

Gianazza, E., Crawford, J. & Miller, I. (2007). Detecting oxidative post-translational modifications in proteins. *Amino Acids* **33**, 51-56.

Gonzalez-Flecha, B. & Demple, B. (1995). Metabolic sources of hydrogen peroxide in aerobically growing *Escherichia coli. J Biol Chem* **270**, 13681-13687.

Graf, J., Meierhofer, R., Wegelin, M. & Mosler, H.-J. (2008). Water disinfection and hygiene behaviour in an urban slum in Kenya: impact on childhood diarrhoea and influence of beliefs. *Int J Environ Health Res* **18**, 335 - 355.

Grant, R. A., Filman, D. J., Finkel, S. E., Kolter, R. & Hogle, J. M. (1998). The crystal structure of Dps, a ferritin homolog that binds and protects DNA. *Nat Struct Biol* **5**, 294-303.

Gross, L. (2007). Paradox Resolved? The Strange Case of the Radiation-Resistant Bacteria. *PLoS Biol* **5**, e108.

Grune, T., Jung, T., Merker, K. & Davies, K. J. (2004). Decreased proteolysis caused by protein aggregates, inclusion bodies, plaques, lipofuscin, ceroid, and 'aggresomes' during oxidative stress, aging, and disease. *Int J Biochem Cell Biol* **36**, 2519-2530.

Hammes, F., Berney, M., Wang, Y., Vital, M., Koster, O. & Egli, T. (2008). Flow-cytometric total bacterial cell counts as a descriptive microbiological parameter for drinking water treatment processes. *Water Res* **42**, 269-277.

Harm, W. (1968). Effects of dose fractionation on ultraviolet survival of *Escherichia coli. Photochem Photobiol* **7**, 73-86.

Harm, W. (1980). *Biological effects of ultraviolet radiation.* Cambridge: Cambridge University Press.

Hartman, P. S. & Eisenstark, A. (1978). Synergistic killing of Escherichia coli by near-UV radiation and hydrogen peroxide: distinction between recA-repairable and recA-nonrepairable damage. *J Bacteriol* **133**, 769-774.

Hartman, P. S., Eisenstark, A. & Pauw, P. G. (1979). Inactivation of phage T7 by near-ultraviolet radiation plus hydrogen peroxide: DNA-protein crosslinks prevent DNA injection. *Proc Natl Acad Sci USA* **76**, 3228-3232.

Heri, S. & Mosler, H. J. (2008). Factors affecting the diffusion of solar water disinfection: A field study in Bolivia. *Health Education & Behavior* **35**, 541-560.

Hobbins, M., Mäusezahl, D. & Tanner, M. (2003). The SODIS health impact study: Swiss Tropical Institute Basel.

Hobbins, M. (2004). Home-based water purification through sunlight: from promotion to health effectiveness. Basel: University of Basel.

Hoefel, D., Grooby, W. L., Monis, P. T., Andrews, S. & Saint, C. P. (2003). Enumeration of water-borne bacteria using viability assays and flow cytometry: a comparison to culture-based techniques. *J Microbiol Methods* **55**, 585-597.

Hoerter, J. D., Arnold, A. A., Kuczynska, D. A. & other authors (2005a). Effects of sublethal UVA irradiation on activity levels of oxidative defence enzymes and protein oxidation in *Escherichia coli*. *J Photochem Photobiol B* **81**, 171-180.

Hoerter, J. D., Arnold, A. A., Ward, C. S., Sauer, M., Johnson, S., Fleming, T. & Eisenstark, A. (2005b). Reduced hydroperoxidase (HPI and HPII) activity in the Deltafur mutant contributes to increased sensitivity to UVA radiation in *Escherichia coli*. *J Photochem Photobiol B* **79**, 151-157.

Hoerter, J. D., Ward, C. S., Bale, K. D. & other authors (2008). Effect of UVA fluence rate on indicators of oxidative stress in human dermal fibroblasts. *Int J of Biol Sci* **4**, 63-70.

Hollaender, A. (1943). Effect of long ultraviolet and short visible radiation (3500 to 4900A) on *Escherichia coli*. *J Bacteriol* **46**, 531-541.

Hunter, P. R. (2009). Household water treatment in developing countries: comparing different intervention types using meta-regression. *Environ Sci Technol* **43**, 8991-8997.

Imlay, J. A. (2009). Oxidative Stress, Module 5.4.4. In *EcoSal*. Edited by J. Foster: ASM Press.

References

Jagger, J. (1972). Growth delay and photoprotection induced by near-ultraviolet light. *Res Prog Org Biol Med Chem* **3 Pt 1**, 383-401.

Jagger, J. (1981). Near-UV radiation effects on microorganisms. *Photochem Photobiol* **34**, 761-768.

Jagger, J. (1985). *Solar-UV actions on living cells*. New York: Praeger Publishers.

Jakob, U., Muse, W., Eser, M. & Bardwell, J. C. (1999). Chaperone activity with a redox switch. *Cell* **96**, 341-352.

Jakob, U., Eser, M. & Bardwell, J. C. (2000). Redox switch of hsp33 has a novel zinc-binding motif. *J Biol Chem* **275**, 38302-38310.

Jeffrey, W. H., Kase, J. P. & Wilhelm, S. W. (2005). UV radiation effects on heterotrophic bacterioplankton and viruses in marine ecosystems. In *The Effects of UV Radiation in the Marine Environment*, pp. 206-236. Edited by S. De Mora, S. Demers & M. Vernet. Cambridge: Cambridge University Press.

Jiang, Y., Rabbi, M., Kim, M., Ke, C., Lee, W., Clark, R. L., Mieczkowski, P. A. & Marszalek, P. E. (2009). UVA generates pyrimidine dimers in DNA directly. *Biophys J* **96**, 1151-1158.

John, R. A. (2002). Photometric assays. In *Enzyme assays*, pp. 49-78. Edited by R. Eisenthal & M. J. Danson. Oxford: Oxford University Press.

Joyce, T. M., McGuigan, K. G., Elmore-Meegan, M. & Conroy, R. M. (1996). Inactivation of fecal bacteria in drinking water by solar heating. *Appl Environ Microbiol* **62**, 399-402.

Kalisvaart, B. F. (2001). Photobiological effects of polychromatic medium pressure UV lamps. *Water Sci Technol* **43**, 191-197.

Kalisvaart, B. F. (2004). Re-use of wastewater: preventing the recovery of pathogens by using medium-pressure UV lamp technology. *Water Sci Technol* **50**, 337-344.

Kapuscinski, R. B. & Mitchell, R. (1981). Solar radiation induces sublethal injury in *Escherichia coli* in seawater. *Appl Environ Microbiol* **41**, 670-674.

Kapuscinski, R. B. (1983). Sunlight-induced mortality of viruses and *Escherichia coli* in costal seawater. *Environ Sci Technol* **17**, 1-6.

Kell, D. B., Kaprelyants, A. S., Weichart, D. H., Harwood, C. R. & Barer, M. R. (1998). Viability and activity in readily culturable bacteria: a review and discussion of the practical issues. *Antonie Van Leeuwenhoek* **73**, 169-187.

Kelland, L. R., Moss, S. H. & Davies, D. J. (1983a). Recovery of *Escherichia coli* K-12 from near-ultraviolet radiation-induced membrane damage. *Photochem Photobiol* **37**, 617-622.

Kelland, L. R., Moss, S. H. & Davies, D. J. (1983b). An action spectrum for ultraviolet radiation-induced membrane damage in *Escherichia coli* K-12. *Photochem Photobiol* **37**, 301-306.

Kelland, L. R., Moss, S. H. & Davies, D. J. (1984). Leakage of 86Rb+ after ultraviolet irradiation of *Escherichia coli* K-12. *Photochem Photobiol* **39**, 329-335.

Keller, A., Nesvizhskii, A. I., Kolker, E. & Aebersold, R. (2002). Empirical statistical model to estimate the accuracy of peptide identifications made by MS/MS and database search. *Anal Chem* **74**, 5383-5392.

Khaengraeng, R. & Reed, R. H. (2005). Oxygen and photoinactivation of *Escherichia coli* in UVA and sunlight. *J Appl Microbiol* **99**, 39-50.

King, B. J., Hoefel, D., Daminato, D. P., Fanok, S. & Monis, P. T. (2008). Solar UV reduces *Cryptosporidium parvum* oocyst infectivity in environmental waters. *J Appl Microbiol* **104**, 1311-1323.

Kirchman, D., K'Nees, E. & Hodson, R. (1985). Leucine incorporation and its potential as a measure of protein synthesis by bacteria in natural aquatic systems. *Appl Environ Microbiol* **49**, 599-607.

Klamen, D. L. & Tuveson, R. W. (1982). The effect of membrane fatty acid composition on the near-UV (300-400 nm) sensitivity of Escherichia coli K1060. *Photochem Photobiol* **35**, 167-173.

Kobayashi, H., Miyamoto, T., Hashimoto, Y., Kiriki, M., Motomatsu, A., Honjoh, K. & Iio, M. (2005). Identification of factors involved in recovery of heat-injured *Salmonella* Enteritidis. *J Food Prot* **68**, 932-941.

Koch, A. L., Doyle, R. J. & Kubitschek, H. E. (1976). Inactivation of membrane transport in *Escherichia coli* by near-ultraviolet light. *J Bacteriol* **126**, 140-146.

Kogure, K., Simidu, U., Taga, N. & Colwell, R. R. (1987). Correlation of direct viable counts with heterotrophic activity for marine bacteria. *Appl Environ Microbiol* **53**, 2332-2337.

References

Kohanski, M. A., Dwyer, D. J., Hayete, B., Lawrence, C. A. & Collins, J. J. (2007). A common mechanism of cellular death induced by bactericidal antibiotics. *Cell* **130**, 797-810.

Kohn, T., Grandbois, M., McNeill, K. & Nelson, K. L. (2007). Association with natural organic matter enhances the sunlight-mediated inactivation of MS2 coliphage by singlet oxygen. *Environ Sci Technol* **41**, 4626-4632.

Kohn, T. & Nelson, K. L. (2007). Sunlight-mediated inactivation of MS2 coliphage via exogenous singlet oxygen produced by sensitizers in natural waters. *Environ Sci Technol* **41**, 192-197.

Komanapalli, I. R., Mudd, J. B. & Lau, B. H. S. (1997). Effect of ozone on metabolic activities of *Escherichia coli* K-12. *Toxicol Letters* **90**, 61-66.

Kramer, G. F., Baker, J. C. & Ames, B. N. (1988). Near-UV stress in *Salmonella typhimurium*: 4-thiouridine in tRNA, ppGpp, and ApppGpp as components of an adaptive response. *J Bacteriol* **170**, 2344-2351.

Kubitschek, H. E. & Doyle, R. J. (1981). Growth delay induced in *Escherichia coli* by near-ultraviolet radiation: relationship to membrane transport functions. *Photochem Photobiol* **33**, 695-702.

Kunert, K. J., Cresswell, C. F., Schmidt, A., Mullineaux, P. M. & Foyer, C. H. (1990). Variations in the activity of glutathione reductase and the cellular glutathione content in relation to sensitivity to methylviologen in *Escherichia coli*. *Arch Biochem Biophys* **282**, 233-238.

Laemmli, U. K. (1970). Cleavage of structural proteins during the assembly of the head of bacteriophage T4. *Nature* **227**, 680-685.

Latch, D. E. & McNeill, K. (2006). Microheterogeneity of singlet oxygen distributions in irradiated humic acid solutions. *Science* **311**, 1743-1747.

Leven, S., Heimberger, A. & Eisenstark, A. (1990). Catalase HPI influences membrane permeability in *Escherichia coli* following near-UV stress. *Biochem Biophys Res Commun* **171**, 1224-1228.

Levine, R. L., Williams, J. A., Stadtman, E. R. & Shacter, E. (1994). Carbonyl assays for determination of oxidatively modified proteins. *Methods Enzymol* **233**, 346-357.

Levine, R. L. & Stadtman, E. R. (2001). Oxidative modification of proteins during aging. *Exp Gerontol* **36**, 1495-1502.

Levine, R. L. (2002). Carbonyl modified proteins in cellular regulation, aging, and disease. *Free Radic Biol Med* **32**, 790-796.

Lindner, A. B., Madden, R., Dernarez, A., Stewart, E. J. & Taddei, F. (2008). Asymmetric segregation of protein aggregates is associated with cellular aging and rejuvenation. *P Natl Acad Sci USA* **105**, 3076-3081.

Long, J., Liu, C., Sun, L., Gao, H. & Liu, J. (2009). Neuronal mitochondrial toxicity of malondialdehyde: inhibitory effects on respiratory function and enzyme activities in rat brain mitochondria. *Neurochem Res* **34**, 786-794.

Lonnen, J., Kilvington, S., Kehoe, S. C., Al-Touati, F. & McGuigan, K. G. (2005). Solar and photocatalytic disinfection of protozoan, fungal and bacterial microbes in drinking water. *Water Res* **39**, 877-883.

Lund, M. N., Luxford, C., Skibsted, L. H. & Davies, M. J. (2008). Oxidation of myosin by haem proteins generates myosin radicals and protein cross-links. *Biochem J* **410**, 565-574.

Maisonneuve, E., Ezraty, B. & Dukan, S. (2008a). Protein aggregates: an aging factor involved in cell death. *J Bacteriol* **190**, 6070-6075.

Maisonneuve, E., Fraysse, L., Lignon, S., Capron, L. & Dukan, S. (2008b). Carbonylated proteins are detectable only in a degradation-resistant aggregate state in *Escherichia coli. J Bacteriol* **190**, 6609-6614.

Maisonneuve, E., Fraysse, L., Moinier, D. & Dukan, S. (2008c). Existence of abnormal protein aggregates in healthy *Escherichia coli* cells. *J Bacteriol* **190**, 887-893.

Maisonneuve, E., Ducret, A., Khoueiry, P., Lignon, S., Longhi, S., Talla, E. & Dukan, S. (2009). Rules governing selective protein carbonylation. *PLoS ONE* **4**, e7269.

Malato, S., Fernández-Ibáñez, P., Maldonado, M. I., Blanco, J. & Gernjak, W. (2009). Decontamination and disinfection of water by solar photocatalysis: Recent overview and trends. *Catalysis Today* **In Press, Corrected Proof**.

Mandal, T. K. & Chatterjee, S. N. (1980). Ultraviolet- and sunlight-induced lipid peroxidation in liposomal membrane. *Radiat Res* **83**, 290-302.

Martin-Dominguez, A., Alarcon-Herrera, M. T., Martin-Dominguez, I. R. & Gonzalez-Herrera, A. (2005). Efficiency in the disinfection of water for human consumption in rural communities using solar radiation. *Sol Energy* **78**, 31-40.

References

Martin, J. W., Chin, J. W. & Nguyen, T. (2003). Reciprocity law experiments in polymeric photodegradation: a critical review. *Prog Org Coat* **47**, 292-311.

Mäusezahl, D., Christen, A., Pacheco, G. D. & other authors (2009). Solar drinking water disinfection (SODIS) to reduce childhood diarrhoea in rural Bolivia: a cluster-randomized, controlled trial. *PLoS Med* **6**, e1000125.

Mazzulli, J. R., Hodara, R., Lind, S. & Ischiropoulos, H. (2006). Oxidative stress and protein deposition diseases. In *Protein misfolding, aggregation and conformational diseases; Part A: Protein aggregation and conformational diseases*. Edited by V. N. Uversky & A. L. Fink: Springer.

McCormick, J. P., Fischer, J. R., Pachlatko, J. P. & Eisenstark, A. (1976). Characterization of a cell-lethal product from the photooxidation of tryptophan: hydrogen peroxide. *Science* **191**, 468-469.

McGuigan, K. G., Joyce, T. M., Conroy, R. M., Gillespie, J. B. & Elmore-Meegan, M. (1998). Solar disinfection of drinking water contained in transparent plastic bottles: characterizing the bacterial inactivation process. *J Appl Microbiol* **84**, 1138-1148.

Mendez-Hermida, F., Ares-Mazas, E., McGuigan, K. G., Boyle, M., Sichel, C. & Fernandez-Ibanez, P. (2007). Disinfection of drinking water contaminated with *Cryptosporidium parvum* oocysts under natural sunlight and using the photocatalyst TiO2. *J Photochem Photobiol B* **88**, 105-111.

Merwald, H., Klosner, G., Kokesch, C., Der-Petrossian, M., Honigsmann, H. & Trautinger, F. (2005). UVA-induced oxidative damage and cytotoxicity depend on the mode of exposure. *J Photochem Photobiol B* **79**, 197-207.

Mirzaei, H. & Regnier, F. (2005). Affinity chromatographic selection of carbonylated proteins followed by identification of oxidation sites using tandem mass spectrometry. *Anal Chem* **77**, 2386-2392.

Mirzaei, H. & Regnier, F. (2006a). Creation of allotypic active sites during oxidative stress. *J Proteome Res* **5**, 2159-2168.

Mirzaei, H. & Regnier, F. (2006b). Protein-RNA cross-linking in the ribosomes of yeast under oxidative stress. *J Proteome Res* **5**, 3249-3259.

Mirzaei, H. & Regnier, F. (2007). Identification of yeast oxidized proteins: chromatographic top-down approach for identification of carbonylated, fragmented and cross-linked proteins in yeast. *J Chromatogr A* **1141**, 22-31.

120

Moriarty, D. J. W. & Bell, R. T. (1993). Bacterial growth and starvation in aquatic environments. In *Starvation in bacteria*, pp. 1-23. Edited by S. Kjelleberg. New York and London: Plenum Press.

Moss, S. H. & Smith, K. C. (1981). Membrane damage can be a significant factor in the inactivation of *Escherichia coli* by near-ultraviolet radiation. *Photochem Photobiol* **33**, 203-210.

Murakami, K., Tsubouchi, R., Fukayama, M., Ogawa, T. & Yoshino, M. (2006). Oxidative inactivation of reduced NADP-generating enzymes in *E. coli*: iron-dependent inactivation with affinity cleavage of NADP-isocitrate dehydrogenase. *Arch Microbiol* **186**, 385-392.

Murphey, W. H., Barrie Kitto, G. & John, M. L. (1969). Malate dehydrogenase from *Escherichia coli*. In *Methods in Enzymology*, pp. 145-147: Academic Press.

Natarajan, A. & Srienc, F. (2000). Glucose uptake rates of single *E. coli* cells grown in glucose-limited chemostat cultures. *J Microbiol Methods* **42**, 87-96.

Neidhardt, F. C., Ingraham, J. L. & Schaechter, M. (1990). *Physiology of the bacterial cell: a molecular approach*. Sunderland, Mass: Sinauer Associates Inc.

Nesvizhskii, A. I., Keller, A., Kolker, E. & Aebersold, R. (2003). A statistical model for identifying proteins by tandem mass spectrometry. *Anal Chem* **75**, 4646-4658.

Nicholls, D. G. & Ferguson, S. (1992). *Bioenergetics 3*. London: Academic Press.

Niki, E., Yoshida, Y., Saito, Y. & Noguchi, N. (2005). Lipid peroxidation: Mechanisms, inhibition, and biological effects. *Biochem Biophys Res Com* **338**, 668-676.

Nyström, T. (2003). Nonculturable bacteria: programmed survival forms or cells at death's door? *BioEssays* **25**, 204-211.

Nyström, T. (2005). Role of oxidative carbonylation in protein quality control and senescence. *EMBO J* **24**, 1311-1317.

Nyström, T. (2006). Oxidative damage and cellular senescence: lessons from bacteria and yeast. In *Redox proteomics: from protein modifications to cellular dysfunction and diseases*, pp. 473-484. Edited by I. Dalle-Donne, A. Scaloni & D. A. Butterfield. New Jersey: John Wiley & Sons.

Nyström, T. (2007). A bacterial kind of aging. *PLoS Genet* **3**, e224.

References

Oates, P. M., Shanahan, P. & Polz, M. F. (2003). Solar disinfection (SODIS): simulation of solar radiation for global assessment and application for point-of-use water treatment in Haiti. *Water Res* **37**, 47-54.

Oguma, K., Katayama, H. & Ohgaki, S. (2002). Photoreactivation of *Escherichia coli* after low- or medium-pressure UV disinfection determined by an endonuclease sensitive site assay. *Appl Environ Microbiol* **68**, 6029-6035.

Oppezzo, O. J. & Pizarro, R. A. (2001). Sublethal effects of ultraviolet A radiation on *Enterobacter cloacae*. *J Photochem Photobiol B* **62**, 158-165.

Pizarro, R. A. & Orce, L. V. (1988). Membrane damage and recovery associated with growth delay induced by near-UV radiation in *Escherichia coli* K-12. *Photochem Photobiol* **47**, 391-397.

Pizarro, R. A. (1995). UV-A oxidative damage modified by environmental conditions in *Escherichia coli*. *Int J Radiat Biol* **68**, 293-299.

Rabilloud, T. (2009). Membrane proteins and proteomics: Love is possible, but so difficult. *Electrophoresis* **30**, S174-S180.

Ramabhadran, T. V. & Jagger, J. (1976). Mechanism of growth delay induced in *Escherichia coli* by near ultraviolet radiation. *Proc Natl Acad Sci USA* **73**, 59-63.

Ravanat, J. L., Douki, T. & Cadet, J. (2001). Direct and indirect effects of UV radiation on DNA and its components. *J Photochem Photobiol B* **63**, 88-102.

Reed, R. H. (1997a). Solar inactivation of faecal bacteria in water : the critical role of oxygen. *Lett Appl Microbiol* **24**, 276-280.

Reed, R. H. (1997b). Solar inactivation of faecal bacteria in water: the critical role of oxygen. *Lett Appl Microbiol* **24**, 276-280.

Reed, R. H., Mani, S. K. & Meyer, V. (2000). Solar photo-oxidative disinfection of drinking water: preliminary field observations. *Lett Appl Microbiol* **30**, 432-436.

Requena, J. R., Chao, C. C., Levine, R. L. & Stadtman, E. R. (2001). Glutamic and aminoadipic semialdehydes are the main carbonyl products of metal-catalyzed oxidation of proteins. *Proc Natl Acad Sci U S A* **98**, 69-74.

Requena, J. R., Levine, R. L. & Stadtman, E. R. (2003). Recent advances in the analysis of oxidized proteins. *Amino Acids* **25**, 221-226.

Rincon, A. G. & Pulgarin, C. (2004). Field solar *E. coli* inactivation in the absence and presence of TiO2: is UV solar dose an appropriate parameter for standardization of water solar disinfection? *Sol Energy* **77**, 635-648.

Robb, F. T., Hauman, J. H. & Peak, M. J. (1978). Similar spectra for the inactivation by monochromatic light of two distinct leucine transport systems in *Escherichia coli. Photochem Photobiol* **27**, 465-469.

Rose, A., Roy, S., Abraham, V. & other authors (2006). Solar disinfection of water for diarrhoeal prevention in southern India. *Arch Dis Child* **91**, 139-141.

Roszak, D. & Colwell, R. R. (1985). Viable but non-culturable bacteria in the aquatic environment. *J Appl Bacteriol* **59**, R9-R9.

Salaj-Smic, E., Dzidic, S. & Trgovcevic, Z. (1985). The effect of a split UV dose on survival, division delay and mutagenesis in *Escherichia coli. Mutat Res Lett* **144**, 127-130.

Sammartano, L. J. & Tuveson, R. W. (1987). *Escherichia coli* strains carrying the cloned cytochrome d terminal oxidase complex are sensitive to near-UV inactivation. *J Bacteriol* **169**, 5304-5307.

Santoni, V., Molloy, M. & Rabilloud, T. (2000). Membrane proteins and proteomics: un amour impossible? *Electrophoresis* **21**, 1054-1070.

Shacter, E., Williams, J. A., Lim, M. & Levine, R. L. (1994). Differential susceptibility of plasma proteins to oxidative modification: examination by western blot immunoassay. *Free Radic Biol Med* **17**, 429-437.

Sharma, R. C. & Jagger, J. (1981). Ultraviolet (254-405 nm) action spectrum and kinetic studies of alanine uptake in *Escherichia coli* B/R. *Photochem Photobiol* **33**, 173-177.

Shevchenko, A., Wilm, M., Vorm, O. & Mann, M. (1996). Mass spectrometric sequencing of proteins silver-stained polyacrylamide gels. *Anal Chem* **68**, 850-858.

Sichel, C., de Cara, M., Tello, J., Blanco, J. & Fernandez-Ibanez, P. (2007). Solar photocatalytic disinfection of agricultural pathogenic fungi: *Fusarium* species. *Appl Catal B-Environ* **74**, 152-160.

Smith, R. J., Kehoe, S. C., McGuigan, K. G. & Barer, M. R. (2000). Effects of simulated solar disinfection of water on infectivity of *Salmonella typhimurium. Lett Appl Microbiol* **31**, 284-288.

Sobsey, M. D., Stauber, C. E., Casanova, L. M., Brown, J. M. & Elliott, M. A. (2008). Point of use household drinking water filtration: a practical, effective solution for providing sustained access to safe drinking water in the developing world. *Environ Sci Technol* **42**, 4261-4267.

References

Sommer, R., Haider, T., Cabaj, A., Heidenreich, E. & Kundi, M. (1996). Increased inactivation of *Saccharomyces cerevisiae* by protraction of UV irradiation. *Appl Environ Microbiol* **62**, 1977-1983.

Sommer, R., Pribil, W., Appelt, S., Gehringer, P., Eschweiler, H., Leth, H., Cabaj, A. & Haider, T. (2001). Inactivation of bacteriophages in water by means of non-ionizing (UV-253.7 nm) and ionizing (gamma) radiation: a comparative approach. *Water Res* **35**, 3109-3116.

Squier, T. C. (2001). Oxidative stress and protein aggregation during biological aging. *Exp Gerontol* **36**, 1539-1550.

Stadtman, E. R. & Levine, R. L. (2003). Free radical-mediated oxidation of free amino acids and amino acid residues in proteins. *Amino Acids* **25**, 207-218.

Stadtman, E. R. (2006). Protein oxidation and aging. *Free Radic Res* **40**, 1250-1258.

Stewart, E. J., Madden, R., Paul, G. & Taddei, F. (2005). Aging and death in an organism that reproduces by morphologically symmetric division. *PLoS Biol* **3**, e45.

Switala, J. & Loewen, P. C. (2002). Diversity of properties among catalases. *Arch Biochem Biophys* **401**, 145-154.

Tamarit, J., Cabiscol, E. & Ros, J. (1998a). Identification of the major oxidatively damaged proteins in *Escherichia coli* cells exposed to oxidative stress. *J Biol Chem* **273**, 3027-3032.

Tamarit, J., Cabiscol, E. & Ros, J. (1998b). Identification of the major oxidatively damaged proteins in *Escherichia coli* cells exposed to oxidative stress. *J Biol Chem* **273**, 3027-3032.

Tarmy, E. M. & Kaplan, N. O. (1968). Chemical characterization of d-lactate dehydrogenase from *Escherichia coli* B. *J Biol Chem* **243**, 2579-2586.

Terman, A. & Brunk, U. T. (2006). Oxidative stress, accumulation of biological 'garbage', and aging. *Antioxid Redox Signal* **8**, 197-204.

Tuveson, R. W., Kagan, J., Shaw, M. A., Moresco, G. M., von Behne, E. M., Pu, H., Bazin, M. & Santus, R. (1987). Phototoxic effects of fluoranthene, a polycyclic aromatic hydrocarbon, on bacterial species. *Environ Mol Mutagen* **10**, 245-261.

Tyrrell, R. (1976). Synergistic lethal action of UV-radiations and mild heat in *Escherichia coli*. *Photochem Photobiol* **24**, 345-351.

Unden, G. & Dünnwald, P. (2008). The aerobic and anaerobic respiratory chain of *Escherichia coli* and *Salmonella enterica*: enzymes and energetics, Module 3.2.2. In *EcoSal.* Edited by V. Stewart: ASM Press.

Varghese, S., Tang, Y. & Imlay, J. A. (2003). Contrasting sensitivities of *Escherichia coli* aconitases A and B to oxidation and iron depletion. *J Bacteriol* **185**, 221-230.

Villarino, A., Bouvet, O. M., Regnault, B., Martin-Delautre, S. & Grimont, P. A. D. (2000). Exploring the frontier between life and death in Escherichia coli: evaluation of different viability markers in live and heat- or UV-killed cells. *Res Microbiol* **151**, 755-768.

Villarino, A., Rager, M. N., Grimont, P. A. & Bouvet, O. M. (2003). Are UV-induced nonculturable Escherichia coli K-12 cells alive or dead? *Eur J Biochem* **270**, 2689-2695.

Visick, J. E. & Clarke, S. (1997). RpoS- and OxyR-independent induction of HPI catalase at stationary phase in *Escherichia coli* and identification of rpoS mutations in common laboratory strains. *J Bacteriol* **179**, 4158-4163.

Vital, M., Fuchslin, H. P., Hammes, F. & Egli, T. (2007). Growth of *Vibrio cholerae* O1 Ogawa Eltor in freshwater. *Microbiology* **153**, 1993-2001.

Vital, M., Hammes, F. & Egli, T. (2008). *Escherichia coli* O157 can grow in natural freshwater at low carbon concentrations. *Environ Microbiol* **10**, 2387-2396.

Voss, P., Hajimiragha, H., Engels, M., Ruhwiedel, C., Calles, C., Schroeder, P. & Grune, T. (2007). Irradiation of GAPDH: a model for environmentally induced protein damage. *Biol Chem* **388**, 583-592.

Wang, Y., Hammes, F., Boon, N. & Egli, T. (2007). Quantification of the filterability of freshwater bacteria through 0.45, 0.22, and 0.1 microm pore size filters and shape-dependent enrichment of filterable bacterial communities. *Environ Sci Technol* **41**, 7080-7086.

Webb, R. B. & Lorenz, J. R. (1970). Oxygen dependence and repair of lethal effects of near ultraviolet and visible light. *Photochem Photobiol* **12**, 283-289.

Webb, R. B. & Brown, M. S. (1979). Oxygen dependence of sensitization to 245-nm radiation by prior exposure to 365-nm radiation in strains of Escherichia coli K12 differing in DNA repair capability. *Radiat Res* **80**, 82-91.

Wegelin, M., Canonica, S., Mechsner, K., Fleischmann, T., Pesaro, F. & Metzler, A. (1994). Solar water disinfection: scope of the process and analysis of radiation experiments. *J Water SRT - Aqua* **43**, 154.

References

Wickner, S., Maurizi, M. R. & Gottesman, S. (1999). Posttranslational quality control: folding, refolding, and degrading proteins. *Science* **286**, 1888-1893.

Winter, J., Linke, K., Jatzek, A. & Jakob, U. (2005). Severe oxidative stress causes inactivation of DnaK and activation of the redox-regulated chaperone Hsp33. *Mol Cell* **17**, 381-392.

Yuanbin, L., Gary, F. & David, S. (2002). Generation of reactive oxygen species by the mitochondrial electron transport chain. *J Neurochem* **80**, 780-787.

Zepp, R. G., Faust, B. C. & Hoigne, J. (1992). Hydroxyl radical formation in aqueous reactions (pH 3-8) of iron (II) with hydrogen-peroxide - the photo-Fenton reaction. *Environ Sci Technol* **26**, 313-319.

Zimmer-Thomas, J. L., Slawson, R. M. & Huck, P. M. (2007). A comparison of DNA repair and survival of *Escherichia coli* O157:H7 following exposure to both low- and medium-pressure UV irradiation. *J Water Health* **5**, 407-415.

www.ingramcontent.com/pod-product-compliance
Lightning Source LLC
Chambersburg PA
CBHW021104210326
41598CB00016B/1322